# 50% OFF Online ACT Prep Course!

Dear Customer,

We consider it an honor and a privilege that you chose our ACT Study Guide. As a way of showing our appreciation and to help us better serve you, we have partnered with Mometrix Test Preparation to offer you **50% off their online ACT Course.** Many ACT courses are needlessly expensive and don't deliver enough value. With their course, you get access to the best ACT prep material, and you only pay half price.

**Mometrix has structured their online course to perfectly complement your printed study guide**. The ACT Online Course contains **in-depth lessons** that cover all the most important topics, **180+ video reviews** that explain difficult concepts, over **1,500 practice questions** to ensure you feel prepared, and more than **500 digital flashcards**, so you can study while you're on the go.

**Online ACT Prep Course**

*Topics Covered:*
- English
  - Organization, Unity, and Cohesion
  - Punctuation and Grammar
- Math
  - Numbers and Operations
  - Algebra and Geometry
- Reading
  - Key Ideas and Details
  - Purpose, Position, and Point of View
- Science
  - Evaluation of Models
  - Interpretation of Data
- Writing

*Course Features:*
- ACT Study Guide
  - Get content that complements our best-selling study guide.
- Full-Length Practice Tests
  - With over 1,500 practice questions, you can test yourself again and again.
- Mobile Friendly
  - If you need to study on the go, the course is easily accessible from your mobile device.
- ACT Flashcards
  - Their course includes a flashcards mode with over 500 content cards for you to study.

To receive this discount, visit their website: mometrix.com/university/act and add the course to your cart. At the checkout page, enter the discount code: **TPBACT50**

If you have any questions or concerns, please contact them at universityhelp@mometrix.com.

Sincerely,

# FREE Test Taking Tips DVD Offer

To help us better serve you, we have developed a Test Taking Tips DVD that we would like to give you for FREE. **This DVD covers world-class test taking tips that you can use to be even more successful when you are taking your test.**

All that we ask is that you email us your feedback about your study guide. Please let us know what you thought about it – whether that is good, bad or indifferent.

To get your **FREE Test Taking Tips DVD**, email freedvd@studyguideteam.com with "FREE DVD" in the subject line and the following information in the body of the email:

       a. The title of your study guide.

       b. Your product rating on a scale of 1-5, with 5 being the highest rating.

       c. Your feedback about the study guide. What did you think of it?

       d. Your full name and shipping address to send your free DVD.

If you have any questions or concerns, please don't hesitate to contact us at freedvd@studyguideteam.com.

Thanks again!

# ACT Math
# Prep Book 2021 and 2022 with
# 3 Mathematics Practice Tests
# [3rd Edition Workbook]

Joshua Rueda

Interested in buying more than 10 copies of our product? Contact us about bulk discounts:
bulkorders@studyguideteam.com

ISBN 13: 9781637750674
ISBN 10: 1637750676

# Table of Contents

# Quick Overview

As you draw closer to taking your exam, effective preparation becomes more and more important. Thankfully, you have this study guide to help you get ready. Use this guide to help keep your studying on track and refer to it often.

This study guide contains several key sections that will help you be successful on your exam. The guide contains tips for what you should do the night before and the day of the test. Also included are test-taking tips. Knowing the right information is not always enough. Many well-prepared test takers struggle with exams. These tips will help equip you to accurately read, assess, and answer test questions.

A large part of the guide is devoted to showing you what content to expect on the exam and to helping you better understand that content. In this guide are practice test questions so that you can see how well you have grasped the content. Then, answer explanations are provided so that you can understand why you missed certain questions.

Don't try to cram the night before you take your exam. This is not a wise strategy for a few reasons. First, your retention of the information will be low. Your time would be better used by reviewing information you already know rather than trying to learn a lot of new information. Second, you will likely become stressed as you try to gain a large amount of knowledge in a short amount of time. Third, you will be depriving yourself of sleep. So be sure to go to bed at a reasonable time the night before. Being well-rested helps you focus and remain calm.

Be sure to eat a substantial breakfast the morning of the exam. If you are taking the exam in the afternoon, be sure to have a good lunch as well. Being hungry is distracting and can make it difficult to focus. You have hopefully spent lots of time preparing for the exam. Don't let an empty stomach get in the way of success!

When travelling to the testing center, leave earlier than needed. That way, you have a buffer in case you experience any delays. This will help you remain calm and will keep you from missing your appointment time at the testing center.

Be sure to pace yourself during the exam. Don't try to rush through the exam. There is no need to risk performing poorly on the exam just so you can leave the testing center early. Allow yourself to use all of the allotted time if needed.

Remain positive while taking the exam even if you feel like you are performing poorly. Thinking about the content you should have mastered will not help you perform better on the exam.

Once the exam is complete, take some time to relax. Even if you feel that you need to take the exam again, you will be well served by some down time before you begin studying again. It's often easier to convince yourself to study if you know that it will come with a reward!

# Test-Taking Strategies

## 1. Predicting the Answer

When you feel confident in your preparation for a multiple-choice test, try predicting the answer before reading the answer choices. This is especially useful on questions that test objective factual knowledge. By predicting the answer before reading the available choices, you eliminate the possibility that you will be distracted or led astray by an incorrect answer choice. You will feel more confident in your selection if you read the question, predict the answer, and then find your prediction among the answer choices. After using this strategy, be sure to still read all of the answer choices carefully and completely. If you feel unprepared, you should not attempt to predict the answers. This would be a waste of time and an opportunity for your mind to wander in the wrong direction.

## 2. Reading the Whole Question

Too often, test takers scan a multiple-choice question, recognize a few familiar words, and immediately jump to the answer choices. Test authors are aware of this common impatience, and they will sometimes prey upon it. For instance, a test author might subtly turn the question into a negative, or he or she might redirect the focus of the question right at the end. The only way to avoid falling into these traps is to read the entirety of the question carefully before reading the answer choices.

## 3. Looking for Wrong Answers

Long and complicated multiple-choice questions can be intimidating. One way to simplify a difficult multiple-choice question is to eliminate all of the answer choices that are clearly wrong. In most sets of answers, there will be at least one selection that can be dismissed right away. If the test is administered on paper, the test taker could draw a line through it to indicate that it may be ignored; otherwise, the test taker will have to perform this operation mentally or on scratch paper. In either case, once the obviously incorrect answers have been eliminated, the remaining choices may be considered. Sometimes identifying the clearly wrong answers will give the test taker some information about the correct answer. For instance, if one of the remaining answer choices is a direct opposite of one of the eliminated answer choices, it may well be the correct answer. The opposite of obviously wrong is obviously right! Of course, this is not always the case. Some answers are obviously incorrect simply because they are irrelevant to the question being asked. Still, identifying and eliminating some incorrect answer choices is a good way to simplify a multiple-choice question.

## 4. Don't Overanalyze

Anxious test takers often overanalyze questions. When you are nervous, your brain will often run wild, causing you to make associations and discover clues that don't actually exist. If you feel that this may be a problem for you, do whatever you can to slow down during the test. Try taking a deep breath or counting to ten. As you read and consider the question, restrict yourself to the particular words used by the author. Avoid thought tangents about what the author *really* meant, or what he or she was *trying* to say. The only things that matter on a multiple-choice test are the words that are actually in the question. You must avoid reading too much into a multiple-choice question, or supposing that the writer meant something other than what he or she wrote.

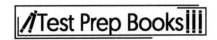

## 5. No Need for Panic

It is wise to learn as many strategies as possible before taking a multiple-choice test, but it is likely that you will come across a few questions for which you simply don't know the answer. In this situation, avoid panicking. Because most multiple-choice tests include dozens of questions, the relative value of a single wrong answer is small. As much as possible, you should compartmentalize each question on a multiple-choice test. In other words, you should not allow your feelings about one question to affect your success on the others. When you find a question that you either don't understand or don't know how to answer, just take a deep breath and do your best. Read the entire question slowly and carefully. Try rephrasing the question a couple of different ways. Then, read all of the answer choices carefully. After eliminating obviously wrong answers, make a selection and move on to the next question.

## 6. Confusing Answer Choices

When working on a difficult multiple-choice question, there may be a tendency to focus on the answer choices that are the easiest to understand. Many people, whether consciously or not, gravitate to the answer choices that require the least concentration, knowledge, and memory. This is a mistake. When you come across an answer choice that is confusing, you should give it extra attention. A question might be confusing because you do not know the subject matter to which it refers. If this is the case, don't eliminate the answer before you have affirmatively settled on another. When you come across an answer choice of this type, set it aside as you look at the remaining choices. If you can confidently assert that one of the other choices is correct, you can leave the confusing answer aside. Otherwise, you will need to take a moment to try to better understand the confusing answer choice. Rephrasing is one way to tease out the sense of a confusing answer choice.

## 7. Your First Instinct

Many people struggle with multiple-choice tests because they overthink the questions. If you have studied sufficiently for the test, you should be prepared to trust your first instinct once you have carefully and completely read the question and all of the answer choices. There is a great deal of research suggesting that the mind can come to the correct conclusion very quickly once it has obtained all of the relevant information. At times, it may seem to you as if your intuition is working faster even than your reasoning mind. This may in fact be true. The knowledge you obtain while studying may be retrieved from your subconscious before you have a chance to work out the associations that support it. Verify your instinct by working out the reasons that it should be trusted.

## 8. Key Words

Many test takers struggle with multiple-choice questions because they have poor reading comprehension skills. Quickly reading and understanding a multiple-choice question requires a mixture of skill and experience. To help with this, try jotting down a few key words and phrases on a piece of scrap paper. Doing this concentrates the process of reading and forces the mind to weigh the relative importance of the question's parts. In selecting words and phrases to write down, the test taker thinks about the question more deeply and carefully. This is especially true for multiple-choice questions that are preceded by a long prompt.

## 9. Subtle Negatives

One of the oldest tricks in the multiple-choice test writer's book is to subtly reverse the meaning of a question with a word like *not* or *except*. If you are not paying attention to each word in the question, you can easily be led astray by this trick. For instance, a common question format is, "Which of the following is...?" Obviously, if the question instead is, "Which of the following is not...?," then the answer will be quite different. Even worse, the test makers are aware of the potential for this mistake and will include one answer choice that would be correct if the question were not negated or reversed. A test taker who misses the reversal will find what he or she believes to be a correct answer and will be so confident that he or she will fail to reread the question and discover the original error. The only way to avoid this is to practice a wide variety of multiple-choice questions and to pay close attention to each and every word.

## 10. Reading Every Answer Choice

It may seem obvious, but you should always read every one of the answer choices! Too many test takers fall into the habit of scanning the question and assuming that they understand the question because they recognize a few key words. From there, they pick the first answer choice that answers the question they believe they have read. Test takers who read all of the answer choices might discover that one of the latter answer choices is actually *more* correct. Moreover, reading all of the answer choices can remind you of facts related to the question that can help you arrive at the correct answer. Sometimes, a misstatement or incorrect detail in one of the latter answer choices will trigger your memory of the subject and will enable you to find the right answer. Failing to read all of the answer choices is like not reading all of the items on a restaurant menu: you might miss out on the perfect choice.

## 11. Spot the Hedges

One of the keys to success on multiple-choice tests is paying close attention to every word. This is never truer than with words like almost, most, some, and sometimes. These words are called "hedges" because they indicate that a statement is not totally true or not true in every place and time. An absolute statement will contain no hedges, but in many subjects, the answers are not always straightforward or absolute. There are always exceptions to the rules in these subjects. For this reason, you should favor those multiple-choice questions that contain hedging language. The presence of qualifying words indicates that the author is taking special care with his or her words, which is certainly important when composing the right answer. After all, there are many ways to be wrong, but there is only one way to be right! For this reason, it is wise to avoid answers that are absolute when taking a multiple-choice test. An absolute answer is one that says things are either all one way or all another. They often include words like *every*, *always*, *best*, and *never*. If you are taking a multiple-choice test in a subject that doesn't lend itself to absolute answers, be on your guard if you see any of these words.

## 12. Long Answers

In many subject areas, the answers are not simple. As already mentioned, the right answer often requires hedges. Another common feature of the answers to a complex or subjective question are qualifying clauses, which are groups of words that subtly modify the meaning of the sentence. If the question or answer choice describes a rule to which there are exceptions or the subject matter is complicated, ambiguous, or confusing, the correct answer will require many words in order to be expressed clearly and accurately. In essence, you should not be deterred by answer choices that seem excessively long. Oftentimes, the author of the text will not be able to write the correct answer without

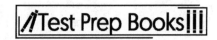

offering some qualifications and modifications. Your job is to read the answer choices thoroughly and completely and to select the one that most accurately and precisely answers the question.

## 13. Restating to Understand

Sometimes, a question on a multiple-choice test is difficult not because of what it asks but because of how it is written. If this is the case, restate the question or answer choice in different words. This process serves a couple of important purposes. First, it forces you to concentrate on the core of the question. In order to rephrase the question accurately, you have to understand it well. Rephrasing the question will concentrate your mind on the key words and ideas. Second, it will present the information to your mind in a fresh way. This process may trigger your memory and render some useful scrap of information picked up while studying.

## 14. True Statements

Sometimes an answer choice will be true in itself, but it does not answer the question. This is one of the main reasons why it is essential to read the question carefully and completely before proceeding to the answer choices. Too often, test takers skip ahead to the answer choices and look for true statements. Having found one of these, they are content to select it without reference to the question above. Obviously, this provides an easy way for test makers to play tricks. The savvy test taker will always read the entire question before turning to the answer choices. Then, having settled on a correct answer choice, he or she will refer to the original question and ensure that the selected answer is relevant. The mistake of choosing a correct-but-irrelevant answer choice is especially common on questions related to specific pieces of objective knowledge. A prepared test taker will have a wealth of factual knowledge at his or her disposal, and should not be careless in its application.

## 15. No Patterns

One of the more dangerous ideas that circulates about multiple-choice tests is that the correct answers tend to fall into patterns. These erroneous ideas range from a belief that B and C are the most common right answers, to the idea that an unprepared test-taker should answer "A-B-A-C-A-D-A-B-A." It cannot be emphasized enough that pattern-seeking of this type is exactly the WRONG way to approach a multiple-choice test. To begin with, it is highly unlikely that the test maker will plot the correct answers according to some predetermined pattern. The questions are scrambled and delivered in a random order. Furthermore, even if the test maker was following a pattern in the assignation of correct answers, there is no reason why the test taker would know which pattern he or she was using. Any attempt to discern a pattern in the answer choices is a waste of time and a distraction from the real work of taking the test. A test taker would be much better served by extra preparation before the test than by reliance on a pattern in the answers.

# FREE DVD OFFER

Don't forget that doing well on your exam includes both understanding the test content and understanding how to use what you know to do well on the test. We offer a completely FREE Test Taking Tips DVD that covers world class test taking tips that you can use to be even more successful when you are taking your test.

All that we ask is that you email us your feedback about your study guide. To get your **FREE Test Taking Tips DVD**, email freedvd@studyguideteam.com with "FREE DVD" in the subject line and the following information in the body of the email:

- The title of your study guide.
- Your product rating on a scale of 1-5, with 5 being the highest rating.
- Your feedback about the study guide. What did you think of it?
- Your full name and shipping address to send your free DVD.

# Introduction to the ACT

## Function of the Test

The ACT is one of two national standardized college entrance examinations (the SAT being the other). Most prospective college students take the ACT or the SAT, and it is increasingly common for students to take both. All four-year colleges and universities in the United States accept the ACT for admissions purposes, and some require it. Some of those schools also use ACT subject scores for placement purposes. Sixteen states also require all high school juniors to take the ACT as part of the states' school evaluation efforts.

The vast majority of people taking the ACT are high school juniors and seniors who intend to apply to college. Traditionally, the SAT was more commonly taken than the ACT, particularly among students on the East and West coasts. However, the popularity of the ACT has grown dramatically in recent years and is now commonly taken by students in all fifty states. In fact, starting in 2013, more test takers took the ACT than the SAT. In 2015, 1.92 million students took the ACT. About 28 percent of 2015 high school graduates taking the ACT met the test's college-readiness benchmarks in all four subjects, while 31 percent met none of the benchmarks.

## Test Administration

The ACT is offered on six dates throughout the year in the U.S. and Canada, and on five of those same dates in other countries. The registration fee includes score reports for four colleges, with additional reports available for purchase. There is a separate registration fee for the optional writing section.

On test dates, the ACT is administered at test centers throughout the world. The test centers are usually high schools or colleges, with several locations usually available in significant population centers.

Test takers can retake the ACT as frequently as the test is offered, up to a maximum of twelve times; although, individual colleges may have limits on how many retakes they will consider. Starting September 2020, test takers can opt to take individual subsections during a retake attempt rather than needing to retake the entire exam.

The ACT does provide reasonable accommodations to test takers with professionally-documented disabilities.

## Test Format

The ACT consists of 215 multiple-choice questions in four subject areas (English, mathematics, reading, and science) and takes about three hours and thirty minutes to complete. It also has an optional writing test, which takes an additional forty minutes.

The English section is 45 minutes long and contains 75 questions on usage, language mechanics, and rhetorical skills. The Mathematics section is 60 minutes long and contains 60 questions on algebra, geometry, and elementary trigonometry. Calculators that meet the ACT's calculator policy are permitted on the Mathematics section. The Reading section is 35 minutes long and contains four written passages with ten questions per passage. The Science section is 35 minutes long and contains 40 questions.

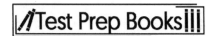

The Writing section is forty minutes long and is always given at the end so that test takers not wishing to take it may leave after completing the other four sections. This section consists of one essay in which students must analyze three different perspectives on a broad social issue. Although the Writing section is optional, some colleges do require it.

| Section | Length | Questions |
|---|---|---|
| English | 45 minutes | 75 |
| Mathematics | 60 minutes | 60 |
| Reading | 35 minutes | 40 |
| Science | 35 minutes | 40 |
| Writing (optional) | 40 minutes | 1 essay |

## Scoring

Test takers receive a score between 1 and 36 for each of the four subject areas. Those scores are averaged together to give a Composite Score, which is the primary score reported as an "ACT score." The most prestigious schools typically admit students with Composite ACT Scores in the low 30's. Other selective schools typically admit students with scores in the high 20's. Traditional colleges more likely admit students with scores in the low 20's, while community colleges and other more open schools typically accept students with scores in the high teens. In 2015, the average composite score among all test takers (including those not applying to college) was 21.

Before September 2020, any test takers who opted to retake the ACT had to retake the entire exam; moreover, the composite score was calculated from the scores achieved over a single testing session. As of September 2020, test takers can select individual sections to retake. Additionally, a "superscore" will be calculated, and will reflect the test taker's highest scores on each of the subsections from any of their attempts at the test. For example, if the test taker took the test twice and earned a higher score on Math and English during the first administration, and a higher score on Reading, Science, and Writing the second time, the "superscore" would be the sum of these top performances.

# ACT Mathematics Test

## What to Expect

The Math test contains 60 multiple-choice questions and lasts 60 minutes. Unlike the English and Reading sections, which each contain four answers options, the math questions all have five choices. Certain calculators are permitted, but the questions are all designed to be answerable without the use of a calculator.

## Tips

- Use your calculator strategically. Some questions are actually solved faster with reasoning or freehand work. Test takers often rely excessively on their calculators and can get bogged down or miss obvious elements of the question. When a calculator is used, consider substituting numbers for variables to solve equations and test your answer.

- Double-check your work. Any extra time in the section can be used performing the reverse operations or plugging your answer into the problem to ensure it is correct.

- Backsolve when you're stuck. If you can't figure out the answer, try using the provided choices and working backward, selecting the one that works. This strategy may not always work and is time-consuming, but it can be helpful with those questions that you cannot figure out.

- Carefully consider all diagrams and figures. Word problems, data interpretation, and geometry questions, in particular, often include important graphics with valuable information. Examine them carefully.

- Consider the reasonableness of your response in the context of the problem. Careless mistakes are often made when test takers are rushing. By evaluating if your response is reasonable, some of these mistakes can be avoided.

## Number and Quantity

### Structure of the Number System

The mathematical number system is made up of two general types of numbers: real and complex. **Real numbers** are those that are used in normal settings, while **complex numbers** are those composed of both a real number and an imaginary one. Imaginary numbers are the result of taking the square root of -1, and $\sqrt{-1} = i$.

The real number system is often explained using a Venn diagram similar to the one below. After a number has been labeled as a real number, further classification occurs when considering the other groups in this diagram. If a number is a never-ending, non-repeating decimal, it falls in the irrational category. Otherwise, it is rational. More information on these types of numbers is provided in the previous section. Furthermore, if a number does not have a fractional part, it is classified as an integer, such as -2, 75, or zero. Whole numbers are an even smaller group that only includes positive integers and zero. The last group of natural numbers is made up of only positive integers, such as 2, 56, or 12.

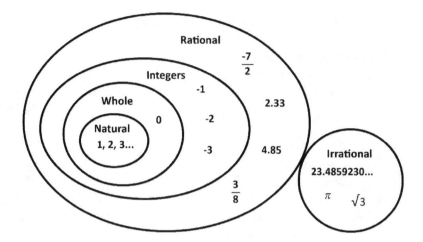

Real numbers can be compared and ordered using the number line. If a number falls to the left on the real number line, it is less than a number on the right. For example, $-2 < 5$ because -2 falls to the left of zero, and 5 falls to the right. Numbers to the left of zero are negative while those to the right are positive.

Complex numbers are made up of the sum of a real number and an imaginary number. Some examples of complex numbers include $6 + 2i$, $5 - 7i$, and $-3 + 12i$. Adding and subtracting complex numbers is similar to collecting like terms. The real numbers are added together, and the imaginary numbers are added together. For example, if the problem asks to simplify the expression $6 + 2i - 3 + 7i$, the 6 and (-3) are combined to make 3, and the $2i$ and $7i$ combine to make $9i$. Multiplying and dividing complex numbers is similar to working with exponents. One rule to remember when multiplying is that $i \times i = -1$. For example, if a problem asks to simplify the expression $4i(3 + 7i)$, the $4i$ should be distributed throughout the 3 and the $7i$. This leaves the final expression $12i - 28$. The 28 is negative because $i \times i$ results in a negative number. The last type of operation to consider with complex numbers is the conjugate. The **conjugate** of a complex number is a technique used to change the complex number into a real number. For example, the conjugate of $4 - 3i$ is $4 + 3i$. Multiplying,

$$(4 - 3i)(4 + 3i)$$

results in $16 + 12i - 12i + 9$,

which has a final answer of $16 + 9 = 25$.

The order of operations—PEMDAS—simplifies longer expressions with real or imaginary numbers. Each operation is listed in the order of how they should be completed in a problem containing more than one operation. Parenthesis can also mean grouping symbols, such as brackets and absolute value. Then, exponents are calculated. Multiplication and division should be completed from left to right, and addition and subtraction should be completed from left to right.

Simplification of another type of expression occurs when radicals are involved. Root is another word for radical. For example, the following expression is a radical that can be simplified: $\sqrt{24x^2}$. First, the number must be factored out to the highest perfect square. Any perfect square can be taken out of a

radical. Twenty-four can be factored into 4 and 6, and 4 can be taken out of the radical. $\sqrt{4} = 2$ can be taken out, and 6 stays underneath. If $x > 0$, $x$ can be taken out of the radical because it is a perfect square. The simplified radical is $2x\sqrt{6}$. An approximation can be found using a calculator.

There are also properties of numbers that are true for certain operations. The **commutative** property allows the order of the terms in an expression to change while keeping the same final answer. Both addition and multiplication can be completed in any order and still obtain the same result. However, order does matter in subtraction and division. The **associative** property allows any terms to be "associated" by parenthesis and retain the same final answer. For example:

$$(4 + 3) + 5 = 4 + (3 + 5)$$

Both addition and multiplication are associative; however, subtraction and division do not hold this property. The **distributive** property states that:

$$a(b + c) = ab + ac$$

It is a property that involves both addition and multiplication, and the $a$ is distributed onto each term inside the parentheses.

## Rational and Irrational Numbers

All real numbers can be separated into two groups: rational and irrational numbers. **Rational numbers** are any numbers that can be written as a fraction, such as $\frac{1}{3}, \frac{7}{4}$, and -25. Alternatively, **irrational numbers** are those that cannot be written as a fraction, such as numbers with never-ending, non-repeating decimal values. Many irrational numbers result from taking roots, such as $\sqrt{2}$ or $\sqrt{3}$. An irrational number may be written as:

34.5684952...

The ellipsis (...) represents the line of numbers after the decimal that does not repeat and is never-ending.

When rational and irrational numbers interact, there are different types of number outcomes. For example, when adding or multiplying two rational numbers, the result is a rational number. No matter what two fractions are added or multiplied together, the result can always be written as a fraction. The following expression shows two rational numbers multiplied together:

$$\frac{3}{8} \times \frac{4}{7} = \frac{12}{56}$$

The product of these two fractions is another fraction that can be simplified to $\frac{3}{14}$.

As another interaction, rational numbers added to irrational numbers will always result in irrational numbers. No part of any fraction can be added to a never-ending, non-repeating decimal to make a rational number. The same result is true when multiplying a rational and irrational number. Taking a fractional part of a never-ending, non-repeating decimal will always result in another never-ending, non-repeating decimal. An example of the product of rational and irrational numbers is shown in the following expression: $2 \times \sqrt{7}$.

The last type of interaction concerns two irrational numbers, where the sum or product may be rational or irrational depending on the numbers being used. The following expression shows a rational sum from two irrational numbers:

$$\sqrt{3} + \left(6 - \sqrt{3}\right) = 6$$

The product of two irrational numbers can be rational or irrational. A rational result can be seen in the following expression:

$$\sqrt{2} \times \sqrt{8} = \sqrt{2 \times 8} = \sqrt{16} = 4$$

An irrational result can be seen in the following:

$$\sqrt{3} \times \sqrt{2} = \sqrt{6}$$

## Integers

An integer is any number that does not have a fractional part. This includes all positive and negative **whole numbers** and zero. Fractions and decimals—which aren't whole numbers—aren't integers.

## Prime Numbers

A **prime** number cannot be divided except by 1 and itself. A prime number has no other factors, which means that no other combination of whole numbers can be multiplied to reach that number. For example, the set of prime numbers between 1 and 27 is {2, 3, 5, 7, 11, 13, 17, 19, 23}.

The number 7 is a prime number because its only factors are 1 and 7. In contrast, 12 isn't a prime number, as it can be divided by other numbers like 2, 3, 4, and 6. Because they are composed of multiple factors, numbers like 12 are called **composite** numbers. All numbers greater than 1 that aren't prime numbers are composite numbers.

## Even and Odd Numbers

An integer is **even** if one of its factors is 2, while those integers without a factor of 2 are **odd**. No numbers except for integers can have either of these labels. For example, 2, 40, -16, and 108 are all even numbers, while -1, 13, 59, and 77 are all odd numbers since they are integers that cannot be divided by 2 without a remainder. Numbers like $0.4$, $\frac{5}{9}$, $\pi$, and $\sqrt{7}$ are neither odd nor even because they are not integers.

## Order of Rational Numbers

A common question type on the ACT asks test takers to order rational numbers from least to greatest or greatest to least. The numbers will come in a variety of formats, including decimals, percentages, roots, fractions, and whole numbers. These questions test for knowledge of different types of numbers and the ability to determine their respective values.

Whether the question asks to order the numbers from greatest to least or least to greatest, the crux of the question is the same—convert the numbers into a common format. Generally, it's easiest to write the numbers as whole numbers and decimals so they can be placed on a number line. The following examples illustrate this strategy:

1. Order the following rational numbers from greatest to least:

$$\sqrt{36}, 0.65, 78\%, \frac{3}{4}, 7, 90\%, \frac{5}{2}$$

Of the seven numbers, the whole number (7) and decimal (0.65) are already in an accessible form, so test takers should concentrate on the other five.

First, the square root of 36 equals 6. (If the test asks for the root of a non-perfect root, determine which two whole numbers the root lies between.) Next, the percentages should be converted to decimals. A percentage means "per hundred," so this conversion requires moving the decimal point two places to the left, leaving 0.78 and 0.9. Lastly, the fractions are evaluated:

$$\frac{3}{4} = \frac{75}{100} = 0.75$$

$$\frac{5}{2} = 2\frac{1}{2} = 2.5$$

Now, the only step left is to list the numbers in the requested order:

$$7, \sqrt{36}, \frac{5}{2}, 90\%, 78\%, \frac{3}{4}, 0.65$$

2. Order the following rational numbers from least to greatest:

$$2.5, \sqrt{9}, -10.5, 0.853, 175\%, \sqrt{4}, \frac{4}{5}$$

$$\sqrt{9} = 3$$

$$175\% = 1.75$$

$$\sqrt{4} = 2$$

$$\frac{4}{5} = 0.8$$

From least to greatest, the answer is: $-10.5, \frac{4}{5}, 0.853, 175\%, \sqrt{4}, 2.5, \sqrt{9}$

## Basic Addition, Subtraction, Multiplication, and Division

Gaining more of something is related to addition, while taking something away relates to subtraction. Vocabulary words such as *total*, *more*, *less*, *left*, and *remain* are common when working with these problems. The $+$ sign means *plus*. This shows that addition is happening. The $-$ sign means *minus*. This shows that subtraction is happening. The symbols will be important when you write out equations.

### Addition
Addition can also be defined in equation form. For example, $4 + 5 = 9$ shows that $4 + 5$ is the same as 9. Therefore, $9 = 9$, and "four plus five equals nine." When two quantities are being added together, the result is called the **sum**. Therefore, the sum of 4 and 5 is 9. The numbers being added, such as 4 and 5, are known as the **addends**.

## Subtraction

Subtraction can also be in equation form. For example, $9 - 5 = 4$ shows that $9 - 5$ is the same as 4 and that "9 minus 5 is 4." The result of subtraction is known as a **difference.** The difference of $9 - 5$ is 4. 4 represents the amount that is left once the subtraction is done. The order in which subtraction is completed does matter. For example, $9 - 5$ and $5 - 9$ do not result in the same answer. $5 - 9$ results in a negative number. So, subtraction does not adhere to the commutative or associative property. The order in which subtraction is completed is important.

## Multiplication

Multiplication is when we add equal amounts. The answer to a multiplication problem is called a **product**. Products stand for the total number of items within different groups. The symbol for multiplication is $\times$ or $\cdot$. We say $2 \times 3$ or $2 \cdot 3$ means "2 times 3."

As an example, there are three sets of four apples. The goal is to know how many apples there are in total. Three sets of four apples gives $4 + 4 + 4 = 12$. Also, three times four apples gives $3 \times 4 = 12$. Therefore, for any whole numbers $a$ and $b$, where $a$ is not equal to zero, $a \times b = b + b + \cdots b$, where $b$ is added $a$ times. Also, $a \times b$ can be thought of as the number of units in a rectangular block consisting of $a$ rows and $b$ columns.

For example, $3 \times 7$ is equal to the number of squares in the following rectangle:

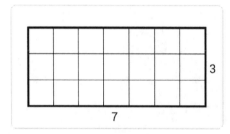

The answer is 21, and there are 21 squares in the rectangle.

When any number is multiplied by one (for example, $8 \times 1 = 8$), the value of original number does not change. Therefore, 1 is the **multiplicative identity**. For any whole number $a$, $1 \times a = a$. Also, any number multiplied by zero results in zero. Therefore, for any whole number $a$, $0 \times a = 0$.

Another method of multiplication can be done with the use of an **area model**. An area model is a rectangle that is divided into rows and columns that match up to the number of place values within each number. For example, $29 \times 65 = 25 + 4$ and $66 = 60 + 5$. The products of those 4 numbers are found within the rectangle and then summed up to get the answer. The entire process is:

$$(60 \times 25) + (5 \times 25) + (60 \times 4) + (5 \times 4)$$

$$1,500 + 240 + 125 + 20 = 1,885$$

Here is the actual area model:

| | 25 | 4 |
|---|---|---|
| **60** | 60x25 <br> 1,500 | 60x4 <br> 240 |
| **5** | 5x25 <br> 125 | 5x4 <br> 20 |

```
        1 , 5 0 0
            2 4 0
            1 2 5
  +            2 0
        1 , 8 8 5
```

## Division

Division is based on dividing a given number into parts. The simplest problem involves dividing a number into equal parts. For example, if a pack of 20 pencils is to be divided among 10 children, you would have to divide 20 by 10. In this example, each child would receive 2 pencils.

The symbol for division is ÷ or /. The equation above is written as $20 \div 10 = 2$, or $20 / 10 = 2$. This means "20 divided by 10 is equal to 2." Division can be explained as the following: for any whole numbers $a$ and $b$, where $b$ is not equal to zero, $a \div b = c$ if—and only if—$a = b \times c$. This means, division can be thought of as a multiplication problem with a missing part. For instance, calculating $20 \div 10$ is the same as asking the following: "If there are 20 items in total with 10 in each group, how many are in each group?" Therefore, 20 is equal to ten times what value? This question is the same as asking, "If there are 20 items in total with 2 in each group, how many groups are there?" The answer to each question is 2.

In a division problem, $a$ is known as the **dividend**, $b$ is the **divisor**, and $c$ is the **quotient**. Zero cannot be divided into parts. Therefore, for any nonzero whole number $a$, $0 \div a = 0$. Also, division by zero is undefined. Dividing an amount into zero parts is not possible.

More difficult division problems involve dividing a number into equal parts, but having some left over. An example is dividing a pack of 20 pencils among 8 friends so that each friend receives the same number of pencils. In this setting, each friend receives 2 pencils, but there are 4 pencils leftover. 20 is the dividend, 8 is the divisor, 2 is the quotient, and 4 is known as the **remainder**. Within this type of division problem, for whole numbers $a$, $b$, $c$, and $d$, $a \div b = c$ with a remainder of $d$. This is true if and only if $a = (b \times c) + d$. When calculating $a \div b$, if there is no remainder, $a$ is said to be *divisible* by $b$. **Even numbers** are all divisible by the number 2. **Odd numbers** are not divisible by 2. An odd number of items cannot be paired up into groups of 2 without having one item leftover.

Dividing a number by a single digit or two digits can be turned into repeated subtraction problems. An area model can be used throughout the problem that represents multiples of the divisor. For example, the answer to $8580 \div 55$ can be found by subtracting 55 from 8580 one at a time and counting the total number of subtractions necessary. However, a simpler process involves using larger multiples of 55.

First, $100 \times 55 = 5,500$ is subtracted from 8,580, and 3,080 is leftover. Next, $50 \times 55 = 2,750$ is subtracted from 3,080 to obtain 380. $5 \times 55 = 275$ is subtracted from 330 to obtain 55, and finally, $1 \times 55 = 55$ is subtracted from 55 to obtain zero. Therefore, there is no remainder, and the answer is:

$$100 + 50 + 5 + 1 = 156$$

Here is a picture of the area model and the repeated subtraction process:

If you want to check the answer of a division problem, multiply the answer by the divisor. This will help you check to see if the dividend is obtained. If there is a remainder, the same process is done, but the remainder is added on at the end to try to match the dividend. In the previous example, $156 \times 55 = 8580$ would be the checking procedure. Dividing decimals involves the same repeated subtraction process. The only difference would be that the subtractions would involve numbers that include values in the decimal places. Lining up decimal places is crucial in this type of problem.

## Order of Operations

When reviewing calculations consisting of more than one operation, the order in which the operations are performed affects the resulting answer. Consider $5 \times 2 + 7$. Performing multiplication then addition results in an answer of 17 because ($5 \times 2 = 10; 10 + 7 = 17$). However, if the problem is written $5 \times (2 + 7)$, the order of operations dictates that the operation inside the parenthesis must be performed first. The resulting answer is 45 because:

$$2 + 7 = 9, \text{so } 5 \times 9 = 45$$

The order in which operations should be performed is remembered using the acronym PEMDAS. PEMDAS stands for parenthesis, exponents, multiplication/division, addition/subtraction. Multiplication and division are performed in the same step, working from left to right with whichever comes first. Addition and subtraction are performed in the same step, working from left to right with whichever comes first.

Consider the following example: $8 \div 4 + 8(7 - 7)$. Performing the operation inside the parenthesis produces $8 \div 4 + 8(0)$ or $8 \div 4 + 8 \times 0$. There are no exponents, so multiplication and division are performed next from left to right resulting in: $2 + 8 \times 0$, then $2 + 0$. Finally, addition and subtraction are performed to obtain an answer of 2. Now consider the following example: $6 \times 3 + 3^2 - 6$.

Parentheses are not applicable. Exponents are evaluated first, which brings us to $6 \times 3 + 9 - 6$. Then multiplication/division forms $18 + 9 - 6$. At last, addition/subtraction leads to the final answer of 21.

## Properties of Operations

Properties of operations exist that make calculations easier and solve problems for missing values. The following table summarizes commonly used properties of real numbers.

| Property | Addition | Multiplication |
|---|---|---|
| Commutative | $a + b = b + a$ | $a \times b = b \times a$ |
| Associative | $(a + b) + c = a + (b + c)$ | $(a \times b) \times c = a \times (b \times c)$ |
| Identity | $a + 0 = a; \ 0 + a = a$ | $a \times 1 = a; \ 1 \times a = a$ |
| Inverse | $a + (-a) = 0$ | $a \times \frac{1}{a} = 1; \ a \neq 0$ |
| Distributive | $a(b + c) = ab + ac$ | |

The **cumulative property of addition** states that the order in which numbers are added does not change the sum. Similarly, the **commutative property of multiplication** states that the order in which numbers are multiplied does not change the product. The **associative property** of addition and multiplication state that the grouping of numbers being added or multiplied does not change the sum or product, respectively. The commutative and associative properties are useful for performing calculations. For example, $(47 + 25) + 3$ is equivalent to $(47 + 3) + 25$, which is easier to calculate.

The **identity property of addition** states that adding zero to any number does not change its value. The **identity property of multiplication** states that multiplying a number by 1 does not change its value. The **inverse property of addition** states that the sum of a number and its opposite equals zero. Opposites are numbers that are the same with different signs (ex. 5 and -5; $-\frac{1}{2}$ and $\frac{1}{2}$). The **inverse property of multiplication** states that the product of a number (other than 0) and its reciprocal equals 1. **Reciprocal numbers** have numerators and denominators that are inverted (ex. $\frac{2}{5}$ and $\frac{5}{2}$). Inverse properties are useful for canceling quantities to find missing values (see algebra content). For example, $a + 7 = 12$ is solved by adding the inverse of 7 (which is -7) to both sides in order to isolate $a$.

The **distributive property** states that multiplying a sum (or difference) by a number produces the same result as multiplying each value in the sum (or difference) by the number and adding (or subtracting) the products. Consider the following scenario: You are buying three tickets for a baseball game. Each ticket costs $18. You are also charged a fee of $2 per ticket for purchasing the tickets online. The cost is calculated: $3 \times 18 + 3 \times 2$. Using the distributive property, the cost can also be calculated:

$$3(18 + 2)$$

## Adding and Subtracting Positive and Negative Numbers

Some problems require adding positive and negative numbers or subtracting positive and negative numbers. Adding a negative number to a positive one can be thought of a reducing or subtracting from the positive number, and the result should be less than the original positive number. For example, adding 8 and -3 is the same is subtracting 3 from 8; the result is 5. This can be visualized by imagining that the positive number (8) represents 8 apples that a student has in her basket. The negative number (-3) indicates the number of apples she is in debt or owes to her friend. In order to pay off her debt and "settle the score," she essentially is in possession of three fewer apples than in her basket ($8 - 3 = 5$), so she actually has five apples that are hers to keep. Should the negative addend be of higher magnitude than the positive addend (for example -9 + 3), the result will be negative, but "less negative" or closer to zero than the large negative number. This is because adding a positive value, even if relatively smaller, to a negative value, reduces the magnitude of the negative in the total. Considering the apple example again, if the girl owed 9 apples to her friend (-9) but she picked 3 (+3) off a tree and gave them to her friend, she now would only owe him six apples (-6), which reduced her debt burden (her negative number of apples) by three.

Subtracting positive and negative numbers works the same way with one key distinction: subtracting a negative number from a negative number yields a "less negative" or more positive result because again, this can be considered as removing or alleviating some debt. For example, if the student with the apples owed 5 apples to her friend, she essentially has -5 applies. If her mom gives that friend 10 apples on behalf of the girl, she now has removed the need to pay back the 5 apples and surpassed neutral (no net apples owed) and now her friend owes *her* five apples (+5).

Stated mathematically: $-5 - (-10) = -5 + 10 = +5$.

When subtracting integers and negative rational numbers, one has to change the problem to adding the opposite and then apply the rules of addition.

- Subtracting two positive numbers is the same as adding one positive and one negative number.

  For example, $4.9 - 7.1$ is the same as $4.9 + (-7.1)$. The solution is -2.2 since the absolute value of -7.1 is greater than 4.9. Another example is $8.5 - 6.4$ which is the same as $8.5 + (-6.4)$. The solution is 2.1 since the absolute value of 8.5 is greater than 6.4.

- Subtracting a positive number from a negative number results in negative value.

  For example, $(-12) - 7$ is the same as $(-12) + (-7)$ with a solution of -19.

- Subtracting a negative number from a positive number results in a positive value.

  For example, $12 - (-7)$ is the same as $12 + 7$ with a solution of 19.

- For multiplication and division of integers and rational numbers, if both numbers are positive or both numbers are negative, the result is a positive value.

  For example, $(-1.7) \times (-4)$ has a solution of 6.8 since both numbers are negative values.

- If one number is positive and another number is negative, the result is a negative value.

  For example, $(-15) \div 5$ has a solution of -3 since there is one negative number.

Adding one positive and one negative number requires taking the absolute values and finding the difference between them. Then, the sign of the number that has the higher absolute value for the final solution is used.

## Operations with Fractions, Decimals, and Percentages

### Fractions

A **fraction** is a part of something that is whole. Items such as apples can be cut into parts to help visualize fractions. If an apple is cut into 2 equal parts, each part represents ½ of the apple. If each half is then cut into two parts, the apple now is cut into quarters. Each piece now represents ¼ of the apple. In this example, each part is equal because they all have the same size. Geometric shapes, such as circles and squares, can also be utilized to help visualize the idea of fractions. For example, a circle can be drawn on the board and divided into 6 equal parts:

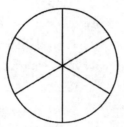

Shading can be used to represent parts of the circle that can be translated into fractions. The top of the fraction, the **numerator,** can represent how many segments are shaded. The bottom of the fraction, the **denominator,** can represent the number of segments that the circle is broken into. A pie is a good analogy to use in this example. If one piece of the circle is shaded, or one piece of pie is cut out, $^1/_6$ of the object is being referred to. An apple, a pie, or a circle can be utilized in order to compare simple fractions. For example, showing that ½ is larger than ¼ and that ¼ is smaller than $^1/_3$ can be accomplished through shading. A **unit fraction** is a fraction in which the numerator is 1, and the denominator is a positive whole number. It represents one part of a whole—one piece of pie.

Imagine that an apple pie has been baked for a holiday party, and the full pie has eight slices. After the party, there are five slices left. How could the amount of the pie that remains be expressed as a fraction? The numerator is 5 since there are 5 pieces left, and the denominator is 8 since there were eight total slices in the whole pie. Thus, expressed as a fraction, the leftover pie totals $\frac{5}{8}$ of the original amount.

Fractions come in three different varieties: proper fractions, improper fractions, and mixed numbers. **Proper fractions** have a numerator less than the denominator, such as $\frac{3}{8}$, but **improper fractions** have a numerator greater than the denominator, such as $\frac{15}{8}$. **Mixed numbers** combine a whole number with a proper fraction, such as $3\frac{1}{2}$. Any mixed number can be written as an improper fraction by multiplying the integer by the denominator, adding the product to the value of the numerator, and dividing the sum by the original denominator. For example:

$$3\frac{1}{2} = \frac{3 \times 2 + 1}{2} = \frac{7}{2}$$

Whole numbers can also be converted into fractions by placing the whole number as the numerator and making the denominator 1. For example, $3 = \frac{3}{1}$.

The bar in a fraction represents division. Therefore $^6/_5$ is the same as $6 \div 5$. In order to rewrite it as a mixed number, division is performed to obtain $6 \div 5 = 1\,R1$. The remainder is then converted into fraction form. The actual remainder becomes the numerator of a fraction, and the divisor becomes the denominator. Therefore $1\,R1$ is written as $1\frac{1}{5}$, a mixed number. A mixed number can also decompose into the addition of a whole number and a fraction. For example,

$$1\frac{1}{5} = 1 + \frac{1}{5} \text{ and } 4\frac{5}{6} = 4 + \frac{1}{6} + \frac{1}{6} + \frac{1}{6} + \frac{1}{6} + \frac{1}{6}$$

Every fraction can be built from a combination of unit fractions.

One of the most fundamental concepts of fractions is their ability to be manipulated by multiplication or division. This is possible since $\frac{n}{n} = 1$ for any non-zero integer. As a result, multiplying or dividing by $\frac{n}{n}$ will not alter the original fraction since any number multiplied or divided by 1 doesn't change the value of that number. Fractions of the same value are known as equivalent fractions. For example, $\frac{2}{8}, \frac{25}{100},$ and $\frac{40}{160}$ are equivalent, as they are all equal to $\frac{1}{4}$.

**Like fractions**, or **equivalent fractions**, are the terms used to describe these fractions that are made up of different numbers but represent the same quantity. For example, the given fractions are $^4/_8$ and $^3/_6$. If a pie was cut into 8 pieces and 4 pieces were removed, half of the pie would remain. Also, if a pie was split into 6 pieces and 3 pieces were eaten, half of the pie would also remain. Therefore, both of the fractions represent half of a pie. These two fractions are referred to as like fractions. **Unlike fractions** are fractions that are different and do not represent equal quantities. When working with fractions in mathematical expressions, like fractions should be simplified. Both $^4/_8$ and $^3/_6$ can be simplified into $^1/_2$.

Comparing fractions can be completed through the use of a number line. For example, if $^3/_5$ and $^6/_{10}$ need to be compared, each fraction should be plotted on a number line. To plot $^3/_5$, the area from 0 to 1 should be broken into 5 equal segments, and the fraction represents 3 of them. To plot $^6/_{10}$, the area from 0 to 1 should be broken into 10 equal segments, and the fraction represents 6 of them.

It can be seen that $\dfrac{3}{5} = \dfrac{6}{10}$

Like fractions are plotted at the same point on a number line. Unit fractions can also be used to compare fractions. For example, if it is known that

$$\frac{4}{5} > \frac{1}{2}$$

and

$$\frac{1}{2} > \frac{4}{10}$$

then it is also known that

$$\frac{4}{5} > \frac{4}{10}$$

Also, converting improper fractions to mixed numbers can be helpful in comparing fractions because the whole number portion of the number is more visible.

Adding and subtracting mixed numbers and fractions can be completed by decomposing fractions into a sum of whole numbers and unit fractions. For example, the given problem is

$$5\frac{3}{7} + 2\frac{1}{7}$$

Decomposing into

$$5 + \frac{1}{7} + \frac{1}{7} + \frac{1}{7} + 2 + \frac{1}{7}$$

This shows that the whole numbers can be added separately from the unit fractions. The answer is:

$$5 + 2 + \frac{1}{7} + \frac{1}{7} + \frac{1}{7} + \frac{1}{7} = 7 + \frac{4}{7} = 7\frac{4}{7}$$

Although many equivalent fractions exist, they are easier to compare and interpret when reduced or simplified. The numerator and denominator of a simple fraction will have no factors in common other than 1. When reducing or simplifying fractions, divide the numerator and denominator by the greatest common factor. A simple strategy is to divide the numerator and denominator by low numbers, like 2, 3, or 5 until arriving at a simple fraction, but the same thing could be achieved by determining the greatest common factor for both the numerator and denominator and dividing each by it. Using the first method is preferable when both the numerator and denominator are even, end in 5, or are obviously a multiple of another number. However, if no numbers seem to work, it will be necessary to factor the numerator and denominator to find the GCF. Let's look at examples:

1) Simplify the fraction $\frac{6}{8}$:

Dividing the numerator and denominator by 2 results in $\frac{3}{4}$, which is a simple fraction.

2) Simplify the fraction $\frac{12}{36}$:

Dividing the numerator and denominator by 2 leaves $\frac{6}{18}$. This isn't a simple fraction, as both the numerator and denominator have factors in common. Dividing each by 3 results in $\frac{2}{6}$, but this can be further simplified by dividing by 2 to get $\frac{1}{3}$. This is the simplest fraction, as the numerator is 1. In cases like this, multiple division operations can be avoided by determining the greatest common factor (12, in this case) between the numerator and denominator.

3) Simplify the fraction $\frac{18}{54}$ by dividing by the greatest common factor:

First, determine the factors for the numerator and denominator. The factors of 18 are 1, 2, 3, 6, 9, and 18. The factors of 54 are 1, 2, 3, 6, 9, 18, 27, and 54. Thus, the greatest common factor is 18. Dividing $\frac{18}{54}$

by 18 leaves $\frac{1}{3}$, which is the simplest fraction. This method takes slightly more work, but it definitively arrives at the simplest fraction.

## *Adding and Subtracting Fractions*

Adding and subtracting fractions that have the same denominators involves adding or subtracting the numerators. The denominator will stay the same. Therefore, the decomposition process can be made simpler, and the fractions do not have to be broken into unit fractions.

For example, the given problem is:

$$4\frac{7}{8} - 2\frac{6}{8}$$

The answer is found by adding the answers to both

$$4 - 2 \text{ and } \frac{7}{8} - \frac{6}{8}$$

$$2 + \frac{1}{8} = 2\frac{1}{8}$$

A common mistake would be to add the denominators so that $\frac{1}{4} + \frac{1}{4} = \frac{1}{8}$ or to add numerators and denominators so that $\frac{1}{4} + \frac{1}{4} = \frac{2}{8}$. However, conceptually, it is known that two quarters make a half, so neither one of these are correct.

If two fractions have different denominators, equivalent fractions must be used to add or subtract them. The fractions must be converted into fractions that have common denominators. A **least common denominator** or the product of the two denominators can be used as the common denominator. For example, in the problem $\frac{5}{6} + \frac{2}{3}$, either 6, which is the least common denominator, or 18, which is the product of the denominators, can be used. In order to use 6, $\frac{2}{3}$ must be converted to sixths. A number line can be used to show the equivalent fraction is $\frac{4}{6}$. What happens is that $\frac{2}{3}$ is multiplied by a fractional form of 1 to obtain a denominator of 6. Hence:

$$\frac{2}{3} \times \frac{2}{2} = \frac{4}{6}$$

Therefore, the problem is now $\frac{5}{6} + \frac{4}{6} = \frac{9}{6}$, which can be simplified into $\frac{3}{2}$. In order to use 18, both fractions must be converted into having 18 as their denominator. $\frac{5}{6}$ would have to be multiplied by $\frac{3}{3}$, and $\frac{2}{3}$ would need to be multiplied by $\frac{6}{6}$. The addition problem would be $\frac{15}{18} + \frac{12}{18} = \frac{27}{18}$, which reduces into $\frac{3}{2}$.

It is always possible to find a common denominator by multiplying the denominators. However, when the denominators are large numbers, this method is unwieldy, especially if the answer must be provided in its simplest form. Thus, it's beneficial to find the **least common denominator** of the fractions—the least common denominator is incidentally also the **least common multiple**.

Once equivalent fractions have been found with common denominators, simply add or subtract the numerators to arrive at the answer:

$$1) \frac{1}{2} + \frac{3}{4} = \frac{2}{4} + \frac{3}{4} = \frac{5}{4}$$

$$2) \frac{3}{12} + \frac{11}{20} = \frac{15}{60} + \frac{33}{60} = \frac{48}{60} = \frac{4}{5}$$

$$3) \frac{7}{9} - \frac{4}{15} = \frac{35}{45} - \frac{12}{45} = \frac{23}{45}$$

$$4) \frac{5}{6} - \frac{7}{18} = \frac{15}{18} - \frac{7}{18} = \frac{8}{18} = \frac{4}{9}$$

## Multiplying and Dividing Fractions

Of the four basic operations that can be performed on fractions, the one that involves the least amount of work is multiplication. To multiply two fractions, simply multiply the numerators together, multiply the denominators together, and place the products of each as a fraction. Whole numbers and mixed numbers can also be expressed as a fraction, as described above, to multiply with a fraction.

Because multiplication is commutative, multiplying a fraction by a whole number is the same as multiplying a whole number by a fraction. The problem involves adding a fraction a specific number of times. The problem $3 \times \frac{1}{4}$ can be translated into adding the unit fraction three times:

$$\frac{1}{4} + \frac{1}{4} + \frac{1}{4} = \frac{3}{4}$$

In the problem $4 \times \frac{2}{5}$, the fraction can be decomposed into $\frac{1}{5} + \frac{1}{5}$ and then added four times to obtain $\frac{8}{5}$. Also, both of these answers can be found by just multiplying the whole number by the numerator of the fraction being multiplied.

The whole numbers can be written in fraction form as:

$$\frac{3}{1} \times \frac{1}{4} = \frac{3}{4}$$

$$\frac{4}{1} \times \frac{2}{5} = \frac{8}{5}$$

Multiplying a fraction times a fraction involves multiplying the numerators together separately and the denominators together separately. For example,

$$\frac{3}{8} \times \frac{2}{3} = \frac{3 \times 2}{8 \times 3} = \frac{6}{24}$$

This can then be reduced to $1/4$.

Dividing a fraction by a fraction is actually a multiplication problem. It involves flipping the divisor and then multiplying normally. For example,

$$\frac{22}{5} \div \frac{1}{2} = \frac{22}{5} \times \frac{2}{1} = \frac{44}{5}$$

The same procedure can be implemented for division problems involving fractions and whole numbers. The whole number can be rewritten as a fraction over a denominator of 1, and then division can be completed.

A common denominator approach can also be used in dividing fractions. Considering the same problem, $\frac{22}{5} \div \frac{1}{2}$, a common denominator between the two fractions is 10. $\frac{22}{5}$ would be rewritten as $\frac{22}{5} \times \frac{2}{2} = \frac{44}{10}$, and $\frac{1}{2}$ would be rewritten as:

$$\frac{1}{2} \times \frac{5}{5} = \frac{5}{10}$$

Dividing both numbers straight across results in:

$$\frac{44}{10} \div \frac{5}{10} = \frac{44/5}{10/10} = \frac{44/5}{1} = 44/5$$

Many real-world problems will involve the use of fractions. Key words include actual fraction values, such as *half, quarter, third, fourth*, etc. The best approach to solving word problems involving fractions is to draw a picture or diagram that represents the scenario being discussed, while deciding which type of operation is necessary in order to solve the problem. A phrase such as "one fourth of 60 pounds of coal" creates a scenario in which multiplication should be used, and the mathematical form of the phrase is $\frac{1}{4} \times 60$.

## Decimals

The **decimal system** is a way of writing out numbers that uses ten different numerals: 0, 1, 2, 3, 4, 5, 6, 7, 8, and 9. This is also called a "base ten" or "base 10" system. Other bases are also used. For example, computers work with a base of 2. This means they only use the numerals 0 and 1.

The **decimal place** denotes how far to the right of the decimal point a numeral is. The first digit to the right of the decimal point is in the **tenths'** place. The next is the **hundredths'** place. The third is the **thousandths'** place.

So, 3.142 has a 1 in the tenths place, a 4 in the hundredths place, and a 2 in the thousandths place.

The **decimal point** is a period used to separate the **ones'** place from the **tenths'** place when writing out a number as a decimal.

A **decimal number** is a number written out with a decimal point instead of as a fraction, for example, 1.25 instead of $\frac{5}{4}$. Depending on the situation, it may be easier to work with fractions, while other times, it may be easier to work with decimal numbers.

A decimal number is **terminating** if it stops at some point. It is called **repeating** if it never stops but repeats a pattern over and over. It is important to note that every rational number can be written as a terminating decimal or as a repeating decimal.

### Addition with Decimals

To add decimal numbers, each number needs to be lined up by the decimal point in vertical columns. For each number being added, the zeros to the right of the last number need to be filled in so that each

of the numbers has the same number of places to the right of the decimal. Then, the columns can be added together. Here is an example of 2.45 + 1.3 + 8.891 written in column form:

$$\begin{array}{r} 2.450 \\ 1.300 \\ + 8.891 \end{array}$$

Zeros have been added in the columns so that each number has the same number of places to the right of the decimal.

Added together, the correct answer is 12.641:

$$\begin{array}{r} 2.450 \\ 1.300 \\ + 8.891 \\ \hline 12.641 \end{array}$$

*Subtraction with Decimals*
Subtracting decimal numbers is the same process as adding decimals. Here is 7.89 − 4.235 written in column form:

$$\begin{array}{r} 7.890 \\ - 4.235 \\ \hline 3.655 \end{array}$$

A zero has been added in the column so that each number has the same number of places to the right of the decimal.

*Multiplication with Decimals*
The simplest way to multiply decimals is to calculate the product as if the decimals are not there, then count the number of decimal places in the original problem. Use that total to place the decimal the same number of places over in your answer, counting from right to left. For example, 0.5 x 1.25 can be rewritten and multiplied as 5 x 125, which equals 625. Then the decimal is added three places from the right for .625.

The final answer will have the same number of decimal places as the total number of decimal places in the problem. The first number has one decimal place, and the second number has two decimal places. Therefore, the final answer will contain three decimal places:

$$0.5 \times 1.25 = 0.625$$

*Division with Decimals*
Dividing a decimal by a whole number entails using long division first by ignoring the decimal point. Then, the decimal point is moved the number of places given in the problem.

For example, 6.8 ÷ 4 can be rewritten as 68 ÷ 4, which is 17. There is one non-zero integer to the right of the decimal point, so the final solution would have one decimal place to the right of the solution. In this case, the solution is 1.7.

Dividing a decimal by another decimal requires changing the divisor to a whole number by moving its decimal point. The decimal place of the dividend should be moved by the same number of places as the divisor. Then, the problem is the same as dividing a decimal by a whole number.

For example, $5.72 \div 1.1$ has a divisor with one decimal point in the denominator. The expression can be rewritten as $57.2 \div 11$ by moving each number one decimal place to the right to eliminate the decimal. The long division can be completed as $572 \div 11$ with a result of 52. Since there is one non-zero integer to the right of the decimal point in the problem, the final solution is 5.2.

In another example, $8 \div 0.16$ has a divisor with two decimal points in the denominator. The expression can be rewritten as $800 \div 16$ by moving each number two decimal places to the right to eliminate the decimal in the divisor. The long division can be completed with a result of 50.

## Percentages

Think of percentages as fractions with a denominator of 100. In fact, **percentage** means "per hundred." Problems often require converting numbers from percentages, fractions, and decimals.

The basic percent equation is the following:

$$\frac{is}{of} = \frac{\%}{100}$$

The placement of numbers in the equation depends on what the question asks.

*Example 1*
Find 40% of 80.

Basically, the problem is asking, "What is 40% of 80?" The 40% is the percent, and 80 is the number to find the percent "of." The equation is:

$$\frac{x}{80} = \frac{40}{100}$$

After cross-multiplying, the problem becomes $100x = 80(40)$. Solving for x gives the answer: $x = 32$.

*Example 2*
What percent of 90 is 20?

The 20 fills in the "is" portion, while 90 fills in the "of." The question asks for the percent, so that will be x, the unknown. The following equation is set up:

$$\frac{20}{90} = \frac{x}{100}$$

Cross-multiplying yields the equation 90x = 20(100). Solving for x gives the answer of 22.2%.

*Example 3*
30% of what number is 30?

The following equation uses the clues and numbers in the problem:

$$\frac{30}{x} = \frac{30}{100}$$

Cross-multiplying results in the equation 30(100) = 30x. Solving for x gives the answer x = 100.

Conversions

*Decimals and Percentages*

Since a percentage is based on "per hundred," decimals and percentages can be converted by multiplying or dividing by 100. Practically speaking, this always involves moving the decimal point two places to the right or left, depending on the conversion. To convert a percentage to a decimal, move the decimal point two places to the left and remove the % sign. To convert a decimal to a percentage, move the decimal point two places to the right and add a "%" sign. Here are some examples:

65% = 0.65
0.33 = 33%
0.215 = 21.5%
99.99% = 0.9999
500% = 5.00
7.55 = 755%

*Fractions and Percentages*

Remember that a percentage is a number per one hundred. So a percentage can be converted to a fraction by making the number in the percentage the numerator and putting 100 as the denominator:

$$43\% = \frac{43}{100}$$

$$97\% = \frac{97}{100}$$

Note that the percent symbol (%) kind of looks like a 0, a 1, and another 0. So think of a percentage like 54% as 54 over 100.

To convert a fraction to a percent, follow the same logic. If the fraction happens to have 100 in the denominator, you're in luck. Just take the numerator and add a percent symbol:

$$\frac{28}{100} = 28\%$$

Otherwise, divide the numerator by the denominator to get a decimal:

$$\frac{9}{12} = 0.75$$

Then convert the decimal to a percentage:

$$0.75 = 75\%$$

Another option is to make the denominator equal to 100. Be sure to multiply the numerator and the denominator by the same number. For example:

$$\frac{3}{20} \times \frac{5}{5} = \frac{15}{100}$$

$$\frac{15}{100} = 15\%$$

*Changing Fractions to Decimals*

To change a fraction into a decimal, divide the denominator into the numerator until there are no remainders. There may be repeating decimals, so rounding is often acceptable. A straight line above the repeating portion denotes that the decimal repeats.

Example: Express 4/5 as a decimal.

Set up the division problem.

$$5\overline{)4}$$

5 does not go into 4, so place the decimal and add a zero.

$$5\overline{)4\,.\,0}$$

5 goes into 40 eight times. There is no remainder.

$$
\begin{array}{r}
0\,.\,8 \\
5\overline{)4\,.\,0} \\
-\,4\,.\,0 \\
\hline
0
\end{array}
$$

The solution is 0.8.

Example: Express 33 1/3 as a decimal.

Since the whole portion of the number is known, set it aside to calculate the decimal from the fraction portion.

Set up the division problem.

$$3\overline{)1}$$

3 does not go into 1, so place the decimal and add zeros. 3 goes into 10 three times.

$$
\begin{array}{r}
0\,.\,3 \\
3\overline{)1\,.\,0}
\end{array}
$$

This will repeat with a remainder of 1.

$$
\begin{array}{r}
0\,.\,3\,3\,3 \\
3\overline{)1\,.\,0\,0\,0} \\
-\,9 \\
\hline
1\,0 \\
-\,9 \\
\hline
1\,0
\end{array}
$$

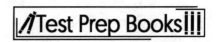

So, we will place a line over the 3 to denote the repetition. The solution is written $0.\overline{3}$.

## Changing Decimals to Fractions

To change decimals to fractions, place the decimal portion of the number—the numerator—over the respective place value—the denominator—then reduce, if possible.

Example: Express 0.25 as a fraction.

This is read as twenty-five hundredths, so put 25 over 100. Then reduce to find the solution.

$$\frac{25}{100} = \frac{1}{4}$$

Example: Express 0.455 as a fraction

This is read as four hundred fifty-five thousandths, so put 455 over 1000. Then reduce to find the solution.

$$\frac{455}{1000} = \frac{91}{200}$$

There are two types of problems that commonly involve percentages. The first is to calculate some percentage of a given quantity, where you convert the percentage to a decimal, and multiply the quantity by that decimal. Secondly, you are given a quantity and told it is a fixed percent of an unknown quantity. In this case, convert to a decimal, then divide the given quantity by that decimal.

Example: What is 30% of 760?

Convert the percent into a useable number. "Of" means to multiply.

$$30\% = 0.30$$

Set up the problem based on the givens, and solve.

$$0.30 \times 760 = 228$$

Example: 8.4 is 20% of what number?

Convert the percent into a useable number.

$$20\% = 0.20$$

The given number is a percent of the answer needed, so divide the given number by this decimal rather than multiplying it.

$$\frac{8.4}{0.20} = 42$$

29

## Factorization

**Factors** are the numbers multiplied to achieve a product. Thus, every product in a multiplication equation has, at minimum, two factors. Of course, some products will have more than two factors. For the sake of most discussions, assume that factors are positive integers.

To find a number's factors, start with 1 and the number itself. Then divide the number by 2, 3, 4, and so on, seeing if any divisors can divide the number without a remainder, keeping a list of those that do. Stop upon reaching either the number itself or another factor.

Let's find the factors of 45. Start with 1 and 45. Then try to divide 45 by 2, which fails. Now divide 45 by 3. The answer is 15, so 3 and 15 are now factors. Dividing by 4 doesn't work, and dividing by 5 leaves 9. Lastly, dividing 45 by 6, 7, and 8 all don't work. The next integer to try is 9, but this is already known to be a factor, so the factorization is complete. The factors of 45 are 1, 3, 5, 9, 15 and 45.

### Prime Factorization

Prime factorization involves an additional step after breaking a number down to its factors: breaking down the factors until they are all prime numbers. A **prime number** is any number that can only be divided by 1 and itself. The prime numbers between 1 and 20 are 2, 3, 5, 7, 11, 13, 17, and 19. As a simple test, numbers that are even or end in 5 are not prime, though there are other numbers that are not prime, but are odd and do not end in 5. For example, 21 is odd and divisible by 1, 3, 7, and 21, so it is not prime.

Let's break 129 down into its prime factors. First, the factors are 3 and 43. Both 3 and 43 are prime numbers, so we're done. But if 43 was not a prime number, then it would also need to be factorized until all of the factors are expressed as prime numbers.

### Common Factor

A **common factor** is a factor shared by two numbers. Let's take 45 and 30 and find the common factors:

> The factors of 45 are: 1, 3, 5, 9, 15, and 45.
> The factors of 30 are: 1, 2, 3, 5, 6, 10, 15, and 30.
> Thus, the common factors are 1, 3, 5, and 15.

### Greatest Common Factor

The **greatest common factor** is the largest number among the shared, common factors. From the factors of 45 and 30, the common factors are 3, 5, and 15. Therefore, 15 is the greatest common factor, as it's the largest number.

### Least Common Multiple

The **least common multiple** is the smallest number that's a multiple of two numbers. Let's try to find the least common multiple of 4 and 9. The multiples of 4 are 4, 8, 12, 16, 20, 24, 28, 32, 36, and so on. For 9, the multiples are 9, 18, 27, 36, 45, 54, etc. Thus, the least common multiple of 4 and 9 is 36 because this is the lowest number where 4 and 9 share multiples.

If two numbers share no factors besides 1 in common, then their least common multiple will be simply their product. If two numbers have common factors, then their least common multiple will be their product divided by their greatest common factor. This can be visualized by the formula $LCM = \frac{x \times y}{GCF}$,

where *x* and *y* are some integers, and *LCM* and *GCF* are their least common multiple and greatest common factor, respectively.

## Exponents

An **exponent** is an operation used as shorthand for a number multiplied or divided by itself for a defined number of times.

$$3^7 = 3 \times 3 \times 3 \times 3 \times 3 \times 3 \times 3$$

In this example, the 3 is called the **base**, and the 7 is called the **exponent**. The exponent is typically expressed as a superscript number near the upper right side of the base but can also be identified as the number following a caret symbol (^). This operation is verbally expressed as "3 to the 7th power" or "3 raised to the power of 7." Common exponents are 2 and 3. A base raised to the power of 2 is referred to as having been "squared," while a base raised to the power of 3 is referred to as having been "cubed."

Several special rules apply to exponents. First, the **Zero Power Rule** finds that any number raised to the zero power equals 1. For example, $100^0$, $2^0$, $(-3)^0$ and $0^0$ all equal 1 because the bases are raised to the zero power.

Second, exponents can be negative. With negative exponents, the equation is expressed as a fraction, as in the following example:

$$3^{-7} = \frac{1}{3^7} = \frac{1}{3 \times 3 \times 3 \times 3 \times 3 \times 3 \times 3}$$

Third, the **Power Rule** concerns exponents being raised by another exponent. When this occurs, the exponents are multiplied by each other:

$$(x^2)^3 = x^6 = (x^3)^2$$

Fourth, when multiplying two exponents with the same base, the **Product Rule** requires that the base remains the same, and the exponents are added. For example, $a^x \times a^y = a^{x+y}$. Since addition and multiplication are commutative, the two terms being multiplied can be in any order.

$$x^3 x^5 = x^{3+5} = x^8 = x^{5+3} = x^5 x^3$$

Fifth, when dividing two exponents with the same base, the **Quotient Rule** requires that the base remains the same, but the exponents are subtracted. So, $a^x \div a^y = a^{x-y}$. Since subtraction and division are not commutative, the two terms must remain in order.

$$x^5 x^{-3} = x^{5-3} = x^2 = x^5 \div x^3 = \frac{x^5}{x^3}$$

Additionally, 1 raised to any power is still equal to 1, and any number raised to the power of 1 is equal to itself. In other words, $a^1 = a$ and $14^1 = 14$.

Exponents play an important role in scientific notation to present extremely large or small numbers as follows: $a \times 10^b$. To write the number in scientific notation, the decimal is moved until there is only one digit on the left side of the decimal point, indicating that the number *a* has a value between 1 and 10. The number of times the decimal moves indicates the exponent to which 10 is raised, here represented by $b$. If the decimal moves to the left, then $b$ is positive, but if the decimal moves to the right, then $b$ is negative. The following examples demonstrate these concepts:

$$3,050 = 3.05 \times 10^3$$

$$-777 = -7.77 \times 10^2$$

$$0.000123 = 1.23 \times 10^{-4}$$

$$-0.0525 = -5.25 \times 10^{-2}$$

## Roots

The **square root** symbol is expressed as $\sqrt{\ }$ and is commonly known as the radical. Taking the root of a number is the inverse operation of multiplying that number by itself some number of times. For example, squaring the number 7 is equal to $7 \times 7$, or 49. Finding the square root is the opposite of finding an exponent, as the operation seeks a number that when multiplied by itself, equals the number in the square root symbol.

For example, $\sqrt{36}$ = 6 because 6 multiplied by 6 equals 36. Note, the square root of 36 is also -6 since -6 × -6 = 36. This can be indicated using a **plus/minus** symbol like this: ±6. However, square roots are often just expressed as a positive number for simplicity, with it being understood that the true value can be either positive or negative.

Perfect squares are numbers with whole number square roots. The list of perfect squares begins with 0, 1, 4, 9, 16, 25, 36, 49, 64, 81, and 100.

Determining the square root of imperfect squares requires a calculator to reach an exact figure. It's possible, however, to approximate the answer by finding the two perfect squares that the number fits between. For example, the square root of 40 is between 6 and 7 since the squares of those numbers are 36 and 49, respectively.

Square roots are the most common root operation. If the radical doesn't have a number to the upper left of the symbol $\sqrt{\ }$, then it's a **square root**. Sometimes a radical includes a number in the upper left, like $\sqrt[3]{27}$, as in the other common root type—the **cube root**. Complicated roots, like the cube root, often require a calculator.

## Estimation

Estimation is finding a value that is close to a solution but is not the exact answer. For example, if there are values in the thousands to be multiplied, then each value can be estimated to the nearest thousand and the calculation performed. This value provides an approximate solution that can be determined very quickly.

**Rounding** is the process of either bumping a number up or leaving it the same, based on a specified place value. First, the place value is specified. Then, the digit to its right is looked at. For example, if rounding to the nearest hundreds place, the digit in the tens place is used. If it is a 0, 1, 2, 3, or 4, the digit being rounded to is left alone. If it is a 5, 6, 7, 8 or 9, the digit being rounded to is increased by one. All other digits before the decimal point are then changed to zeros, and the digits in decimal places are dropped. If a decimal place is being rounded to, all subsequent digits are just dropped. For example, if

845,231.45 was to be rounded to the nearest thousands place, the answer would be 845,000. The 5 would remain the same due to the 2 in the hundreds place. Also, if 4.567 was to be rounded to the nearest tenths place, the answer would be 4.6. The 5 increased to 6 due to the 6 in the hundredths place, and the rest of the decimal is dropped.

Sometimes when performing operations such as multiplying numbers, the result can be estimated by rounding. For example, to estimate the value of $11.2 \times 2.01$, each number can be rounded to the nearest integer. This will yield a result of 22.

Rounding numbers helps with estimation because it changes the given number to a simpler, although less accurate, number than the exact given number. Rounding allows for easier calculations, which estimate the results of using the exact given number. The accuracy of the estimate and ease of use depends on the place value to which the number is rounded. Rounding numbers consists of:

- Determining what place value the number is being rounded to
- Examining the digit to the right of the desired place value to decide whether to round up or keep the digit, and
- Replacing all digits to the right of the desired place value with zeros.

To round 746,311 to the nearest ten thousand, the digit in the ten thousands place should be located first. In this case, this digit is 4 (7**4**6,311). Then, the digit to its right is examined. If this digit is 5 or greater, the number will be rounded up by increasing the digit in the desired place by one. If the digit to the right of the place value being rounded is 4 or less, the number will be kept the same. For the given example, the digit being examined is a 6, which means that the number will be rounded up by increasing the digit to the left by one. Therefore, the digit 4 is changed to a 5. Finally, to write the rounded number, any digits to the left of the place value being rounded remain the same and any to its right are replaced with zeros. For the given example, rounding 746,311 to the nearest ten thousand will produce 750,000. To round 746,311 to the nearest hundred, the digit to the right of the three in the hundreds place is examined to determine whether to round up or keep the same number. In this case, that digit is a 1, so the number will be kept the same and any digits to its right will be replaced with zeros. The resulting rounded number is 746,300.

Rounding place values to the right of the decimal follows the same procedure, but digits being replaced by zeros can simply be dropped. To round 3.752891 to the nearest thousandth, the desired place value is located (3.75**2**891) and the digit to the right is examined. In this case, the digit 8 indicates that the number will be rounded up, and the 2 in the thousandths place will increase to a 3. Rounding up and replacing the digits to the right of the thousandths place produces 3.753000 which is equivalent to 3.753. Therefore, the zeros are not necessary, and the rounded number should be written as 3.753.

When rounding up, if the digit to be increased is a 9, the digit to its left is increased by 1 and the digit in the desired place value is changed to a zero. For example, the number 1,598 rounded to the nearest ten is 1,600. Another example shows the number 43.72961 rounded to the nearest thousandth is 43.730 or 43.73.

## Vectors

A **vector** can be thought of as an abstract list of numbers or as giving a location in a space. For example, the coordinates $(x, y)$ for points in the Cartesian plane are vectors. Each entry in a vector can be referred to by its location in the list: first, second, third, and so on. The total length of the list is the **dimension** of the vector. A vector is often denoted as such by putting an arrow on top of it, e.g.

$$\vec{v} = (v_1, v_2, v_3)$$

## Adding Vectors Graphically and Algebraically

There are two basic operations for vectors. First, two vectors can be added together. Let:

$$\vec{v} = (v_1, v_2, v_3)$$
$$\vec{w} = (w_1, w_2, w_3)$$

The sum of the two vectors is defined to be:

$$\vec{v} + \vec{w} = (v_1 + w_1, v_2 + w_2, v_3 + w_3)$$

Subtraction of vectors can be defined similarly.

Vector addition can be visualized in the following manner. First, each vector can be visualized as an arrow. Then, the base of one arrow is placed at the tip of the other arrow. The tip of this first arrow now hits some point in space, and there will be an arrow from the origin to this point. This new arrow corresponds to the new vector. In subtraction, the direction of the arrow being subtracted is reversed.

For example, if adding together the vectors (-2, 3) and (4, 1), the new vector will be (-2 + 4, 3 + 1), or (2, 4). Graphically, this may be pictured in the following manner:

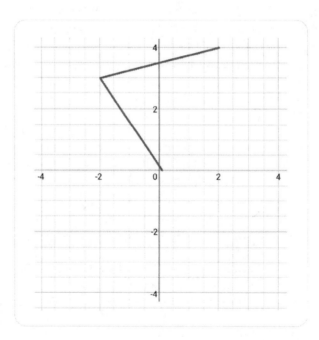

## Performing Scalar Multiplications

The second basic operation for vectors is called **scalar multiplication**. Scalar multiplication is multiplying any vector by any real number, which is denoted here as a scalar. Let $\vec{v} = (v_1, v_2, v_3)$, and let $a$ be an arbitrary real number. Then the scalar multiple $a\vec{v} = (av_1, av_2, av_3)$. Graphically, this corresponds to

changing the length of the arrow corresponding to the vector by a factor, or scale, of $a$. That is why the real number is called a **scalar** in this instance.

As an example, let $\vec{v} = (2, -1, 1)$. Then $3\vec{v} = (3 \cdot 2, 3(-1), 3 \cdot 1) = (6, -3, 3)$.

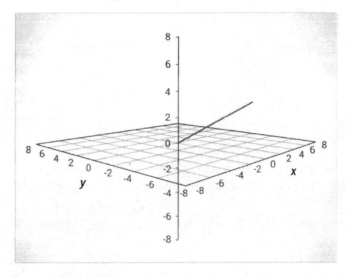

Note that scalar multiplication is **distributive** over vector addition, meaning that $a(\vec{v} + \vec{w}) = a\vec{v} + a\vec{w}$.

## Determinants

A **matrix** is a rectangular arrangement of numbers in rows and columns. The **determinant** of a matrix is a special value that can be calculated for any square matrix.

Using the *square 2 x 2 matrix* $\begin{bmatrix} a & b \\ c & d \end{bmatrix}$, the determinant is $ad - bc$.

For example, the determinant of the matrix $\begin{bmatrix} -5 & 1 \\ 3 & 4 \end{bmatrix}$ is:

$$-5(4) - 1(3) = -20 - 3 = -23$$

Using a *3 x 3 matrix* $\begin{bmatrix} a & b & c \\ d & e & f \\ g & h & i \end{bmatrix}$, the determinant is:

$$a(ei - fh) - b(di - fg) + c(dh - eg)$$

For example, the determinant of the matrix $\begin{bmatrix} 2 & 0 & 1 \\ -1 & 3 & 2 \\ 2 & -2 & -1 \end{bmatrix}$ is

$$2\big(3(-1) - 2(-2)\big) - 0\big(-1(-1) - 2(2)\big) + 1\big(-1(-2) - 3(2)\big)$$

$$2(-3 + 4) - 0(1 - 4) + 1(2 - 6)$$

$$2(1) - 0(-3) + 1(-4)$$

$$2 - 0 - 4 = -2$$

The pattern continues for larger square matrices.

# *Algebra*

## Algebraic Expressions and Equations

An **algebraic expression** is a statement about an unknown quantity expressed in mathematical symbols. A **variable** is used to represent the unknown quantity, usually denoted by a letter. An equation is a statement in which two expressions (at least one containing a variable) are equal to each other. An algebraic expression can be thought of as a mathematical phrase and an equation can be thought of as a mathematical sentence.

Algebraic expressions and equations both contain numbers, variables, and mathematical operations. The following are examples of algebraic expressions: $5x + 3$, $7xy - 8(x^2 + y)$, and $\sqrt{a^2 + b^2}$. An expression can be simplified or evaluated for given values of variables. The following are examples of equations: $2x + 3 = 7$, $a^2 + b^2 = c^2$, and $2x + 5 = 3x - 2$. An equation contains two sides separated by an equal sign. Equations can be solved to determine the value(s) of the variable for which the statement is true.

## Parts of Expressions

Algebraic expressions consist of variables, numbers, and operations. A **term** of an expression is any combination of numbers and/or variables, and terms are separated by addition and subtraction. For example, the expression $5x^2 - 3xy + 4 - 2$ consists of 4 terms: $5x^2$, -3xy, 4y, and -2. Note that each term includes its given sign (+ or −). The **variable** part of a term is a letter that represents an unknown quantity. The **coefficient** of a term is the number by which the variable is multiplied. For the term 4y, the variable is y, and the coefficient is 4. Terms are identified by the power (or exponent) of its variable.

A number without a variable is referred to as a **constant**. If the variable is to the first power ($x^1$ or simply x), it is referred to as a linear term. A term with a variable to the second power ($x^2$) is quadratic, and a term to the third power ($x^3$) is cubic. Consider the expression $x^3 + 3x - 1$. The constant is -1. The linear term is 3x. There is no quadratic term. The cubic term is $x^3$.

An algebraic expression can also be classified by how many terms exist in the expression. Any like terms should be combined before classifying. A **monomial** is an expression consisting of only one term. Examples of monomials are: 17, 2x, and $-5ab^2$. A **binomial** is an expression consisting of two terms separated by addition or subtraction. Examples include $2x - 4$ and $-3y^2 + 2y$. A **trinomial** consists of 3 terms. For example, $5x^2 - 2x + 1$ is a trinomial.

## Adding and Subtracting Linear Algebraic Expressions

An algebraic expression is simplified by combining like terms. As mentioned, term is a number, variable, or product of a number and variables separated by addition and subtraction. For the algebraic expression $3x^2 - 4x + 5 - 5x^2 + x - 3$, the terms are $3x^2$, -4x, 5, -5x^2, x, and -3. Like terms have the same variables raised to the same powers (exponents). The like terms for the previous example are $3x^2$ and -5x^2, -4x and x, 5 and -3. To combine like terms, the coefficients (numerical factor of the term including sign) are added, and the variables and their powers are kept the same. Note that if a coefficient is not written, it is an implied coefficient of 1 ($x = 1x$). The previous example will simplify to:

$$-2x^2 - 3x + 2$$

When adding or subtracting algebraic expressions, each expression is written in parenthesis. The negative sign is distributed when necessary, and like terms are combined. Consider the following:

add $2a + 5b - 2$ to $a - 2b + 8c - 4$

The sum is set as follows:

$$(a - 2b + 8c - 4) + (2a + 5b - 2)$$

In front of each set of parentheses is an implied positive one, which, when distributed, does not change any of the terms. Therefore, the parentheses are dropped and like terms are combined:

$$a - 2b + 8c - 4 + 2a + 5b - 2$$

$$3a + 3b + 8c - 6$$

Consider the following problem: Subtract $2a + 5b - 2$ from $a - 2b + 8c - 4$. The difference is set as follows:

$$(a - 2b + 8c - 4) - (2a + 5b - 2)$$

The implied one in front of the first set of parentheses will not change those four terms. However, distributing the implied -1 in front of the second set of parentheses will change the sign of each of those three terms:

$$a - 2b + 8c - 4 - 2a - 5b + 2$$

Combining like terms yields the simplified expression: $-a - 7b + 8c - 2$.

## Distributive Property

The distributive property states that multiplying a sum (or difference) by a number produces the same result as multiplying each value in the sum (or difference) by the number and adding (or subtracting) the products. Using mathematical symbols, the distributive property states $a(b + c) = ab + ac$. The expression $4(3 + 2)$ is simplified using the order of operations. Simplifying inside the parenthesis first produces $4 \times 5$, which equals 20. The expression $4(3 + 2)$ can also be simplified using the distributive property:

$$4(3 + 2)$$

$$4 \times 3 + 4 \times 2$$

$$12 + 8 = 20$$

Consider the following example: $4(3x - 2)$. The expression cannot be simplified inside the parenthesis because $3x$ and -2 are not like terms and therefore cannot be combined. However, the expression can be simplified by using the distributive property and multiplying each term inside of the parenthesis by the term outside of the parenthesis: $12x - 8$. The resulting equivalent expression contains no like terms, so it cannot be further simplified.

Consider the expression:

$$(3x + 2y + 1) - (5x - 3) + 2(3y + 4)$$

Again, there are no like terms, but the distributive property is used to simplify the expression. Note there is an implied one in front of the first set of parentheses and an implied -1 in front of the second set of parentheses.

Distributing the 1, -1, and 2 produces:

$$1(3x) + 1(2y) + 1(1) - 1(5x) - 1(-3) + 2(3y) + 2(4)$$

$$3x + 2y + 1 - 5x + 3 + 6y + 8$$

This expression contains like terms that are combined to produce the simplified expression:

$$-2x + 8y + 12$$

Algebraic expressions are tested to be equivalent by choosing values for the variables and evaluating both expressions. For example, $4(3x - 2)$ and $12x - 8$ are tested by substituting 3 for the variable $x$ and calculating to determine if equivalent values result.

## Evaluating Expressions for Given Values

An algebraic expression is a statement written in mathematical symbols, typically including one or more unknown values represented by variables. For example, the expression $2x + 3$ states that an unknown number ($x$) is multiplied by 2 and added to 3. If given a value for the unknown number, or variable, the value of the expression is determined. For example, if the value of the variable $x$ is 4, the value of the expression 4 is multiplied by 2, and 3 is added. This results in a value of 11 for the expression.

When given an algebraic expression and values for the variable(s), the expression is evaluated to determine its numerical value. To evaluate the expression, the given values for the variables are substituted (or replaced), and the expression is simplified using the order of operations. Parenthesis should be used when substituting. Consider the following: Evaluate $a - 2b + ab$ for $a = 3$ and $b = -1$. To evaluate, any variable $a$ is replaced with 3 and any variable $b$ with -1, producing:

$$(3) - 2(-1) + (3)(-1)$$

Next, the order of operations is used to calculate the value of the expression, which is 2.

## Verbal Statements and Algebraic Expressions

As mentioned, an algebraic expression is a statement about unknown quantities expressed in mathematical symbols. The statement *five times a number added to forty* is expressed as $5x + 40$. An equation is a statement in which two expressions (with at least one containing a variable) are equal to one another. The statement *five times a number added to forty is equal to ten* is expressed as:

$$5x + 40 = 10$$

Real world scenarios can also be expressed mathematically. Suppose a job pays its employees $300 per week and $40 for each sale made. The weekly pay is represented by the expression $40x + 300$ where $x$ is the number of sales made during the week.

Consider the following scenario: Bob had $20 and Tom had $4. After selling 4 ice cream cones to Bob, Tom has as much money as Bob. The cost of an ice cream cone is an unknown quantity and can be represented by a variable ($x$). The amount of money Bob has after his purchase is four times the cost of

an ice cream cone subtracted from his original $20 $\rightarrow 20 - 4x$. The amount of money Tom has after his sale is four times the cost of an ice cream cone added to his original $4 $\rightarrow 4x + 4$. After the sale, the amount of money that Bob and Tom have is equal $\rightarrow 20 - 4x = 4x + 4$. Solving for $x$ yields $x = 2$.

## Use of Formulas

**Formulas** are mathematical expressions that define the value of one quantity, given the value of one or more different quantities. Formulas look like equations because they contain variables, numbers, operators, and an equal sign. All formulas are equations, but not all equations are formulas. A formula must have more than one variable. For example, $2x + 7 = y$ is an equation and a formula (it relates the unknown quantities $x$ and $y$). However, $2x + 7 = 3$ is an equation but not a formula (it only expresses the value of the unknown quantity $x$).

Formulas are typically written with one variable alone (or isolated) on one side of the equal sign. This variable can be thought of as the *subject* in that the formula is stating the value of the *subject* in terms of the relationship between the other variables. Consider the distance formula: $distance = rate \times time$ or $d = rt$. The value of the subject variable $d$ (distance) is the product of the variable $r$ and $t$ (rate and time). Given the rate and time, the distance traveled can easily be determined by substituting the values into the formula and evaluating.

The formula $P = 2l + 2w$ expresses how to calculate the perimeter of a rectangle ($P$) given its length ($l$) and width ($w$). To find the perimeter of a rectangle with a length of 3ft and a width of 2ft, these values are substituted into the formula for $l$ and $w$:

$$P = 2(3ft) + 2(2ft)$$

Following the order of operations, the perimeter is determined to be 10ft. When working with formulas such as these, including units is an important step.

Given a formula expressed in terms of one variable, the formula can be manipulated to express the relationship in terms of any other variable. In other words, the formula can be rearranged to change which variable is the **subject.** To solve for a variable of interest by manipulating a formula, the equation may be solved as if all other variables were numbers. The same steps for solving are followed, leaving operations in terms of the variables instead of calculating numerical values. For the formula $P = 2l + 2w$, the perimeter is the subject expressed in terms of the length and width. To write a formula to calculate the width of a rectangle, given its length and perimeter, the previous formula relating the three variables is solved for the variable $w$. If $P$ and $l$ were numerical values, this is a two-step linear equation solved by subtraction and division. To solve the equation $P = 2l + 2w$ for $w$, $2l$ is first subtracted from both sides: $P - 2l = 2w$. Then both sides are divided by 2:

$$\frac{P - 2l}{2} = w$$

## Word Problems

Word problems can appear daunting, but prepared test takers shouldn't let the verbiage psyche them out. No matter the scenario or specifics, the key to answering them is to translate the words into a math problem. It is critical to keep in mind what the question is asking and what operations could lead to that answer. The following word problem resembles one of the question types most frequently encountered on the exam.

## Working with Money

Walter's Coffee Shop sells a variety of drinks and breakfast treats.

| Price List | |
|---|---|
| Hot Coffee | $2.00 |
| Slow Drip Iced Coffee | $3.00 |
| Latte | $4.00 |
| Muffins | $2.00 |
| Crepe | $4.00 |
| Egg Sandwich | $5.00 |

| Costs | |
|---|---|
| Hot Coffee | $0.25 |
| Slow Drip Iced Coffee | $0.75 |
| Latte | $1.00 |
| Muffins | $1.00 |
| Crepe | $2.00 |
| Egg Sandwich | $3.00 |

Walter's utilities, rent, and labor costs him $500 per day. Today, Walter sold 200 hot coffees, 100 slow drip iced coffees, 50 lattes, 75 muffins, 45 crepes, and 60 egg sandwiches. What was Walter's total profit today?

To accurately answer this type of question, the first step is to determine the total cost of making his drinks and treats, then determine how much revenue he earned from selling those products. After arriving at these two totals, the profit is measured by deducting the total cost from the total revenue.

Walter's costs for today:

| | | |
|---|---|---|
| 200 hot coffees | × $0.25 | = $50 |
| 100 slow drip iced coffees | × $0.75 | = $75 |
| 50 lattes | × $1.00 | = $50 |
| 75 muffins | × $1.00 | = $75 |
| 45 crepes | × $2.00 | = $90 |
| 60 egg sandwiches | × $3.00 | = $180 |
| Utilities, Rent, and Labor | | = $500 |
| Total costs | | = $1,020 |

Walter's revenue for today:

| | | |
|---|---|---|
| 200 hot coffees | × $2.00 | = $400 |
| 100 slow drip iced coffees | × $3.00 | = $300 |
| 50 lattes | × $4.00 | = $200 |
| 75 muffins | × $2.00 | = $150 |
| 45 crepes | × $4.00 | = $180 |
| 60 egg sandwiches | × $5.00 | = $300 |
| Total revenue | | = $1,530 |

Walter's $Profit = Revenue - Costs = \$1,530 - \$1,020 = \$510$

This strategy can be applied to other question types. For example, calculating salary after deductions, balancing a checkbook, and calculating a dinner bill are common word problems similar to business planning. In all cases, the most important step is remembering to use the correct operations. When a balance is increased, addition is used. When a balance is decreased, the problem requires subtraction. Common sense and organization are one's greatest assets when answering word problems.

## Unit Rate

**Unit rate** word problems ask test takers to calculate the rate or quantity of something in a different value. For example, a problem might say that a car drove a certain number of miles in a certain number of minutes and then ask how many miles per hour the car was traveling. These questions involve solving proportions. Consider the following examples:

1. Alexandra made $96 during the first 3 hours of her shift as a temporary worker at a law office. She will continue to earn money at this rate until she finishes in 5 more hours. How much does Alexandra make per hour? How much money will Alexandra have made at the end of the day?

This problem can be solved in two ways. The first is to set up a proportion, as the rate of pay is constant. The second is to determine her hourly rate, multiply the 5 hours by that rate, and then adding the $96.

To set up a proportion, the money already earned (numerator) is placed over the hours already worked (denominator) on one side of an equation. The other side has $x$ over 8 hours (the total hours worked in the day). It looks like this: $\frac{96}{3} = \frac{x}{8}$. Now, cross-multiply yields $768 = 3x$. To get $x$, the 768 is divided by 3, which leaves $x = 256$. Alternatively, as $x$ is the numerator of one of the proportions, multiplying by its denominator will reduce the solution by one step. Thus, Alexandra will make $256 at the end of the day. To calculate her hourly rate, the total is divided by 8, giving $32 per hour.

Alternatively, it is possible to figure out the hourly rate by dividing $96 by 3 hours to get $32 per hour. Now her total pay can be figured by multiplying $32 per hour by 8 hours, which comes out to $256.

2. Jonathan is reading a novel. So far, he has read 215 of the 335 total pages. It takes Jonathan 25 minutes to read 10 pages, and the rate is constant. How long does it take Jonathan to read one page? How much longer will it take him to finish the novel? Express the answer in time.

To calculate how long it takes Jonathan to read one page, 25 minutes is divided by 10 pages to determine the page per minute rate. Thus, it takes 2.5 minutes to read one page.

Jonathan must read 120 more pages to complete the novel. (This is calculated by subtracting the pages already read from the total.) Now, his rate per page is multiplied by the number of pages. Thus, $120 \times 2.5 = 300$. Expressed in time, 300 minutes is equal to 5 hours.

3. At a hotel, $\frac{4}{5}$ of the 120 rooms are booked for Saturday. On Sunday, $\frac{3}{4}$ of the rooms are booked. On which day are more of the rooms booked, and by how many more?

The first step is to calculate the number of rooms booked for each day. This is done by multiplying the fraction of the rooms booked by the total number of rooms.

$$\text{Saturday:} \frac{4}{5} \times 120 = \frac{4}{5} \times \frac{120}{1} = \frac{480}{5} = 96 \text{ rooms}$$

$$\text{Sunday:} \frac{3}{4} \times 120 = \frac{3}{4} \times \frac{120}{1} = \frac{360}{4} = 90 \text{ rooms}$$

Thus, more rooms were booked on Saturday by 6 rooms.

4. In a veterinary hospital, the veterinarian-to-pet ratio is 1:9. The ratio is always constant. If there are 45 pets in the hospital, how many veterinarians are currently in the veterinary hospital?

A proportion is set up to solve for the number of veterinarians: $\frac{1}{9} = \frac{x}{45}$

Cross-multiplying results in $9x = 45$, which works out to 5 veterinarians.

Alternatively, as there are always 9 times as many pets as veterinarians, it is possible to divide the number of pets (45) by 9. This also arrives at the correct answer of 5 veterinarians.

5. At a general practice law firm, 30% of the lawyers work solely on tort cases. If 9 lawyers work solely on tort cases, how many lawyers work at the firm?

The first step is to solve for the total number of lawyers working at the firm, which will be represented here with $x$. The problem states that 9 lawyers work solely on torts cases, and they make up 30% of the total lawyers at the firm. Thus, 30% multiplied by the total, $x$, will equal 9. Written as equation, this is:

$$30\% \times x = 9$$

It's easier to deal with the equation after converting the percentage to a decimal, leaving $0.3x = 9$. Thus, $x = \frac{9}{0.3} = 30$ lawyers working at the firm.

6. Xavier was hospitalized with pneumonia. He was originally given 35mg of antibiotics. Later, after his condition continued to worsen, Xavier's dosage was increased to 60mg. What was the percent increase of the antibiotics? Round the percentage to the nearest tenth.

An increase or decrease in percentage can be calculated by dividing the difference in amounts by the original amount and multiplying by 100. Written as an equation, the formula is:

$$\frac{new\ quantity\ -\ old\ quantity}{old\ quantity} \times 100$$

Here, the question states that the dosage was increased from 35mg to 60mg, so these values are plugged into the formula to find the percentage increase.

$$\frac{60-35}{35} \times 100 = \frac{25}{35} \times 100 = .7142 \times 100 = 71.4\%$$

## Linear Expressions or Equations in One Variable

Linear expressions and equations are concise mathematical statements that can be written to model a variety of scenarios. Questions found pertaining to this topic will contain one variable only. A variable is an unknown quantity, usually denoted by a letter ($x$, $n$, $p$, etc.). In the case of linear expressions and equations, the power of the variable (its exponent) is 1. A variable without a visible exponent is raised to the first power.

### Writing Linear Expressions and Equations

A linear expression is a statement about an unknown quantity expressed in mathematical symbols. The statement "five times a number added to forty" can be expressed as $5x + 40$. A linear equation is a statement in which two expressions (at least one containing a variable) are equal to each other. The statement "five times a number added to forty is equal to ten" can be expressed as $5x + 40 = 10$. Real-world scenarios can also be expressed mathematically. Consider the following:

Bob had $20 and Tom had $4. After selling 4 ice cream cones to Bob, Tom has as much money as Bob.

The cost of an ice cream cone is an unknown quantity and can be represented by a variable. The amount of money Bob has after his purchase is four times the cost of an ice cream cone subtracted from his original $20. The amount of money Tom has after his sale is four times the cost of an ice cream cone added to his original $4. This can be expressed as: $20 - 4x = 4x + 4$, where $x$ represents the cost of an ice cream cone.

When expressing a verbal or written statement mathematically, it is key to understand words or phrases that can be represented with symbols. The following are examples:

| Symbol | Phrase |
|---|---|
| $+$ | added to, increased by, sum of, more than |
| $-$ | decreased by, difference between, less than, take away |
| $x$ | multiplied by, 3 (4, 5 . . .) times as large, product of |
| $\div$ | divided by, quotient of, half (third, etc.) of |
| $=$ | is, the same as, results in, as much as |
| $x, t, n, etc.$ | a number, unknown quantity, value of |

## Solving Linear Equations

When asked to solve a linear equation, one must determine a numerical value for the unknown variable. Given a linear equation involving addition, subtraction, multiplication, and division, isolation of the variable is done by working backward. Addition and subtraction are inverse operations, as are multiplication and division; therefore, they can be used to cancel each other out.

The first steps to solving linear equations are to distribute if necessary and combine any like terms that are on the same side of the equation. Sides of an equation are separated by an $=$ sign. Next, the equation should be manipulated to get the variable on one side. Whatever is done to one side of an equation, must be done to the other side to remain equal. Then, the variable should be isolated by using inverse operations to undo the order of operations backward. Undo addition and subtraction, then undo multiplication and division. For example:

Solve $4(t - 2) + 2t - 4 = 2(9 - 2t)$

Distribute: $4t - 8 + 2t - 4 = 18 - 4t$

Combine like terms: $6t - 12 = 18 - 4t$

Add 4t to each side to move the variable: $10t - 12 = 18$

Add 12 to each side to isolate the variable: $10t = 30$

Divide each side by 10 to isolate the variable: $t = 3$

The answer can be checked by substituting the value for the variable into the original equation and ensuring both sides calculate to be equal.

## Linear Inequalities in One Variable

Linear inequalities and linear equations are both comparisons of two algebraic expressions. However, unlike equations in which the expressions are equal to each other, linear inequalities compare expressions that are unequal. Linear equations typically have one value for the variable that makes the statement true. Linear inequalities generally have an infinite number of values that make the statement true. Exceptions to these last two statements are covered later on.

## Writing Linear Inequalities

Linear inequalities are a concise mathematical way to express the relationship between unequal values. More specifically, they describe in what way the values are unequal. A value could be greater than ($>$); less than ($<$); greater than or equal to ($\geq$); or less than or equal to ($\leq$) another value. The statement "five times a number added to forty is more than sixty-five" can be expressed as $5x + 40 > 65$. Common words and phrases that express inequalities are:

| Symbol | Phrase |
|--------|--------|
| $<$ | is under, is below, smaller than, beneath |
| $>$ | is above, is over, bigger than, exceeds |
| $\leq$ | no more than, at most, maximum |
| $\geq$ | no less than, at least, minimum |

## Solving Linear Inequalities

When solving a linear inequality, the solution is the set of all numbers that makes the statement true. The inequality $x + 2 \geq 6$ has a solution set of 4 and every number greater than 4 (4.0001, 5, 12, 107, etc.). Adding 2 to 4 or any number greater than 4 would result in a value that is greater than or equal to 6. Therefore, $x \geq 4$ would be the solution set.

Solution sets for linear inequalities often will be displayed using a number line. If a value is included in the set ($\geq$ or $\leq$), there is a shaded dot placed on that value and an arrow extending in the direction of the solutions. For a variable $>$ or $\geq$ a number, the arrow would point right on the number line (the direction where the numbers increase); and if a variable is $<$ or $\leq$ a number, the arrow would point left (where the numbers decrease). If the value is not included in the set ($>$ or $<$), an open circle on that value would be used with an arrow in the appropriate direction.

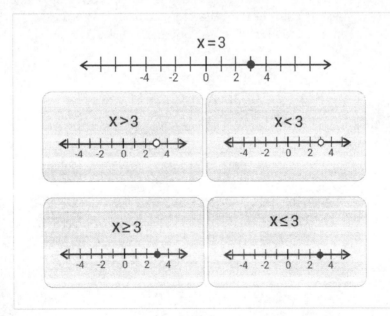

Students may be asked to write a linear inequality given a graph of its solution set. To do so, they should identify whether the value is included (shaded dot or open circle) and the direction in which the arrow is pointing.

In order to algebraically solve a linear inequality, the same steps should be followed as in solving a linear equation (see section on *Solving Linear Equations*). The inequality symbol stays the same for all operations EXCEPT when multiplying or dividing by a negative number. If multiplying or dividing by a negative number while solving an inequality, the relationship reverses (the sign flips). Multiplying or dividing by a positive does not change the relationship, so the sign stays the same. In other words, $>$ switches to $<$ and vice versa. An example is shown below.

Solve $-2(x + 4) \leq 22$ for the value of $x$.

First, distribute -2 to the binomial by multiplying:

$$-2x - 8 \leq 22$$

Next, add 8 to both sides to isolate the variable:

$$-2x \leq 30$$

Divide both sides by -2 to solve for $x$:

$$x \geq -15$$

## Quadratic Equations

A **quadratic equation** is an equation in the form $ax^2 + bx + c = 0$. There are several methods to solve such equations. The easiest method will depend on the quadratic equation in question.

Sometimes, it is possible to solve quadratic equations by manually *factoring* them. This means rewriting them in the form $(x + A)(x + B) = 0$. If this is done, then they can be solved by remembering that when $ab = 0$, either $a$ or $b$ must be equal to zero. Therefore, to have $(x + A)(x + B) = 0$, $(x + A) = 0$ or $(x + B) = 0$ is needed. These equations have the solutions $x = -A$ and $x = -B$, respectively.

In order to factor a quadratic equation, note that:

$$(x + A)(x + B) = x^2 + (A + B)x + AB$$

So, if an equation is in the form $x^2 + bx + c$, two numbers, $A$ and $B$, need to be found that will add up to give us $b$, and multiply together to give us $c$.

As an example, consider solving the equation:

$$-3x^2 + 6x + 9 = 0$$

Start by dividing both sides by $-3$, leaving:

$$x^2 - 2x - 3 = 0$$

Now, notice that $1 - 3 = -2$, and also that $(1)(-3) = -3$. This means the equation can be factored into $(x + 1)(x - 3) = 0$. Now, solve $(x + 1) = 0$ and $(x - 3) = 0$ to get $x = -1$ and $x = 3$ as the solutions.

It is useful when trying to factor to remember these three things:

$$x^2 + 2xy + y^2 = (x + y)^2$$

$$x^2 - 2xy + y^2 = (x - y)^2$$

and $x^2 - y^2 = (x + y)(x - y)$.

However, factoring by hand is often hard to do. If there are no obvious ways to factor the quadratic equation, solutions can still be found by using the *quadratic formula*.

The quadratic formula is:

$$x = \frac{-b \pm \sqrt{b^2 - 4ac}}{2a}$$

This method will always work, although it sometimes can take longer than factoring by hand, if the factors are easy to guess. Using the standard form $ax^2 + bx + c = 0$, plug the values of $a$, $b$, and $c$ from the equation into the formula and solve for x. There will either be two answers, one answer, or no real answer. No real answer comes when the value of the discriminant, the number under the square root, is a negative number. Since there are no real numbers that square to get a negative, the answer will be no real roots.

Here is an example of solving a quadratic equation using the quadratic formula. Suppose the equation to solve is $-2x^2 + 3x + 1 = 0$. There is no obvious way to factor this, so the quadratic formula is used, with $a = -2, b = 3, c = 1$. After substituting these values into the quadratic formula, it yields this:

$$x = \frac{-3 \pm \sqrt{3^2 - 4(-2)(1)}}{2(-2)}$$

This can be simplified to obtain:

$$\frac{3 \pm \sqrt{9 + 8}}{4}$$

or

$$\frac{3 \pm \sqrt{17}}{4}$$

Challenges can be encountered when asked to find a quadratic equation with specific roots. Given roots $A$ and $B$, a quadratic function can be constructed with those roots by taking $(x - A)(x - B)$. So, in constructing a quadratic equation with roots $x = -2, 3$, it would result in:

$$(x + 2)(x - 3) = x^2 - x - 6$$

Multiplying this by a constant also could be done without changing the roots.

## Rewriting Expressions

**Algebraic expressions** are made up of numbers, variables, and combinations of the two, using mathematical operations. Expressions can be rewritten based on their factors. For example, the

47

expression $6x + 4$ can be rewritten as $2(3x + 2)$ because 2 is a factor of both $6x$ and 4. More complex expressions can also be rewritten based on their factors. The expression $x^4 - 16$ can be rewritten as $(x^2 - 4)(x^2 + 4)$. This is a different type of factoring, where a difference of squares is factored into a sum and difference of the same two terms. With some expressions, the factoring process is simple and only leads to a different way to represent the expression. With others, factoring and rewriting the expression leads to more information about the given problem.

In the following quadratic equation, factoring the binomial leads to finding the zeros of the function:

$$x^2 - 5x + 6 = y$$

This equations factors into $(x - 3)(x - 2) = y$, where 2 and 3 are found to be the zeros of the function when y is set equal to zero. The zeros of any function are the x-values where the graph of the function on the coordinate plane crosses the x-axis.

Factoring an equation is a simple way to rewrite the equation and find the zeros, but factoring is not possible for every quadratic. Completing the square is one way to find zeros when factoring is not an option. The following equation cannot be factored:

$$x^2 + 10x - 9 = 0$$

The first step in this method is to move the constant to the right side of the equation, making it $x^2 + 10x = 9$. Then, the coefficient of x is divided by 2 and squared. This number is then added to both sides of the equation, to make the equation still true. For this example, $\left(\frac{10}{2}\right)^2 = 25$ is added to both sides of the equation to obtain:

$$x^2 + 10x + 25 = 9 + 25$$

This expression simplifies to $x^2 + 10x + 25 = 34$, which can then be factored into $(x + 5)^2 = 34$. Solving for x then involves taking the square root of both sides and subtracting 5.

This leads to two zeros of the function:

$$x = \pm\sqrt{34} - 5$$

Depending on the type of answer the question seeks, a calculator may be used to find exact numbers.

Given a **quadratic equation in standard form**— $ax^2 + bx + c = 0$—the sign of $a$ tells whether the function has a minimum value or a maximum value. If $a > 0$, the graph opens up and has a minimum value. If $a < 0$, the graph opens down and has a maximum value. Depending on the way the quadratic equation is written, multiplication may need to occur before a max/min value is determined.

Exponential expressions can also be rewritten, just as quadratic equations. Properties of exponents must be understood. Multiplying two exponential expressions with the same base involves adding the exponents:

$$a^m a^n = a^{m+n}$$

Dividing two exponential expressions with the same base involves subtracting the exponents:

$$\frac{a^m}{a^n} = a^{m-n}$$

Raising an exponential expression to another exponent includes multiplying the exponents:

$$(a^m)^n = a^{mn}$$

The zero power always gives a value of 1: $a^0 = 1$. Raising either a product or a fraction to a power involves distributing that power:

$$(ab)^m = a^m b^m \text{ and } \left(\frac{a}{b}\right)^m = \frac{a^m}{b^m}$$

Finally, raising a number to a negative exponent is equivalent to the reciprocal including the positive exponent:

$$a^{-m} = \frac{1}{a^m}$$

## Polynomials

An expression of the form $ax^n$, where $n$ is a non-negative integer, is called a **monomial** because it contains one term. A sum of monomials is called a **polynomial.** For example, $-4x^3 + x$ is a polynomial, while $5x^7$ is a monomial. A function equal to a polynomial is called a **polynomial function.**

The monomials in a polynomial are also called the **terms** of the polynomial.

The constants that precede the variables are called **coefficients.**

The highest value of the exponent of $x$ in a polynomial is called the **degree** of the polynomial. So, $-4x^3 + x$ has a degree of 3, while $-2x^5 + x^3 + 4x + 1$ has a degree of 5. When multiplying polynomials, the degree of the result will be the sum of the degrees of the two polynomials being multiplied.

Addition and subtraction operations can be performed on polynomials with like terms. **Like terms** refers to terms that have the same variable and exponent. The two following polynomials can be added together by collecting like terms:

$$(x^2 + 3x - 4) + (4x^2 - 7x + 8)$$

The $x^2$ terms can be added as:

$$x^2 + 4x^2 = 5x^2$$

The $x$ terms can be added as $3x + -7x = -4x$, and the constants can be added as $-4 + 8 = 4$. The following expression is the result of the addition:

$$5x^2 - 4x + 4$$

When subtracting polynomials, the same steps are followed, only subtracting like terms together.

Multiplication of polynomials can also be performed. Given the two polynomials, $(y^3 - 4)$ and $(x^2 + 8x - 7)$, each term in the first polynomial must be multiplied by each term in the second polynomial. The steps to multiply each term in the given example are as follows:

$$(y^3 \times x^2) + (y^3 \times 8x) + (y^3 \times -7) + (-4 \times x^2) + (-4 \times 8x) + (-4 \times -7)$$

Simplifying each multiplied part, yields:

$$x^2y^3 + 8xy^3 - 7y^3 - 4x^2 - 32x + 28$$

None of the terms can be combined because there are no like terms in the final expression. Any polynomials can be multiplied by each other by following the same set of steps, then collecting like terms at the end.

## FOIL Method

FOIL is a technique for generating polynomials through the multiplication of binomials. A **polynomial** is an expression of multiple variables (for example, $x, y, z$) in at least three terms involving only the four basic operations and exponents. FOIL is an acronym for First, Outer, Inner, and Last. "First" represents the multiplication of the terms appearing first in the binomials. "Outer" means multiplying the outermost terms. "Inner" means multiplying the terms inside. "Last" means multiplying the last terms of each binomial.

After completing FOIL and solving the operations, **like terms** are combined. To identify like terms, test takers should look for terms with the same variable and the same exponent. For example, in $4x^2 - x^2 + 15x + 2x^2 - 8$, the $4x^2, -x^2$, and $2x^2$ are all like terms because they have the variable ($x$) and exponent (2). Thus, after combining the like terms, the polynomial has been simplified to:

$$5x^2 + 15x - 8$$

The purpose of FOIL is to simplify an equation involving multiple variables and operations. Although it sounds complicated, working through some examples will provide some clarity:

1. Simplify $(x + 10)(x + 4) =$

$(x \times x)$ + $(x \times 4)$ + $(10 \times x)$ + $(10 \times 4)$
First   Outer   Inner   Last

After multiplying these binomials, it's time to solve the operations and combine like terms. Thus, the expression becomes:

$$2x^2 + 4x + 10x + 40 = 2x^2 + 14x + 40$$

2. Simplify $2x(4x^3 - 7y^2 + 3x^2 + 4)$

Here, a monomial ($2x$) is multiplied into a polynomial ($4x^3 - 7y^2 + 3x^2 + 4$). Using the distributive property, the monomial gets multiplied by each term in the polynomial. This becomes:

$$2x(4x^3) - 2x(7y^2) + 2x(3x^2) + 2x(4)$$

Now, each monomial is simplified, starting with the coefficients:

$$(2 \times 4)(x \times x^3) - (2 \times 7)(x \times y^2) + (2 \times 3)(x \times x^2) + (2 \times 4)(x)$$

When multiplying powers with the same base, their exponents are added. Remember, a variable with no listed exponent has an exponent of 1, and exponents of distinct variables cannot be combined. This produces the answer:

$$8x^{1+3} - 14xy^2 + 6x^{1+2} + 8x = 8x^4 - 14xy^2 + 6x^3 + 8x$$

3. Simplify $(8x^{10}y^2z^4) \div (4x^2y^4z^7)$

The first step is to divide the coefficients of the first two polynomials: $8 \div 4 = 2$. The second step is to divide exponents with the same variable, which requires subtracting the exponents. This results in:

$$2(x^{10-2}y^{2-4}z^{4-7}) = 2x^8y^{-2}z^{-3}$$

However, the most simplified answer should include only positive exponents. Thus, $y^{-2}z^{-3}$ needs to be converted into fractions, respectively $\frac{1}{y^2}$ and $\frac{1}{z^3}$. Since the $2x^8$ has a positive exponent, it is placed in the numerator, and $\frac{1}{y^2}$ and $\frac{1}{z^3}$ are combined into the denominator, leaving $\frac{2x^8}{y^2z^3}$ as the final answer.

## Zeros of Polynomials

Finding the **zeros of polynomial functions** is the same process as finding the solutions of polynomial equations. These are the points at which the graph of the function crosses the x-axis. As stated previously, factors can be used to find the zeros of a polynomial function. The degree of the function shows the number of possible zeros. If the highest exponent on the independent variable is 4, then the degree is 4, and the number of possible zeros is 4. If there are complex solutions, the number of roots is less than the degree.

Given the function $y = x^2 + 7x + 6$, $y$ can be set equal to zero, and the polynomial can be factored. The equation turns into $0 = (x + 1)(x + 6)$, where $x = -1$ and $x = -6$ are the zeros. Since this is a quadratic equation, the shape of the graph will be a parabola. Knowing that zeros represent the points where the parabola crosses the x-axis, the maximum or minimum point is the only other piece needed to sketch a rough graph of the function. By looking at the function in standard form, the coefficient of $x$

is positive; therefore, the parabola opens up. Using the zeros and the minimum, the following rough sketch of the graph can be constructed:

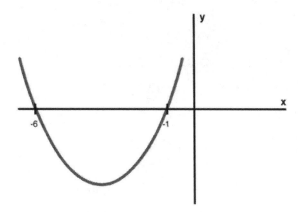

## Rational Expressions and Equations

A **rational expression** is a fraction where the numerator and denominator are both polynomials. Some examples of rational expressions include the following: $\frac{4x^3y^5}{3z^4}$, $\frac{4x^3+3x}{x^2}$, and $\frac{x^2+7x+1}{x+2}$. Since these refer to expressions and not equations, they can be simplified but not solved. Using the rules in the previous *Exponents* and *Roots* sections, some rational expressions with monomials can be simplified. Other rational expressions such as the last example, $\frac{x^2+7x+10}{x+2}$, require more steps to be simplified. First, the polynomial on top can be factored from $x^2 + 7x + 10$ into $(x + 5)(x + 2)$. Then the common factors can be canceled and the expression can be simplified to $(x + 5)$.

The following problem is an example of using rational expressions:

Reggie wants to lay sod in his rectangular backyard. The length of the yard is given by the expression $4x + 2$, and the width is unknown. The area of the yard is $20x + 10$. Reggie needs to find the width of the yard. Knowing that the area of a rectangle is length multiplied by width, an expression can be written to find the width: $\frac{20x+}{4x+2}$, area divided by length. Simplifying this expression by factoring out 10 on the top and 2 on the bottom leads to this expression:

$$\frac{10(2x + 1)}{2(2x + 1)}$$

Canceling out the $2x + 1$ results in $\frac{10}{2} = 5$. The width of the yard is found to be 5 by simplifying the rational expression.

A **rational equation** can be as simple as an equation with a ratio of polynomials, $\frac{p(x)}{q(x)}$, set equal to a value, where $p(x)$ and $q(x)$ are both polynomials. A rational equation has an equal sign, which is different from expressions. This leads to solutions, or numbers that make the equation true.

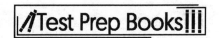

It is possible to solve rational equations by trying to get all of the $x$ terms out of the denominator and then isolating them on one side of the equation. For example, to solve the equation $\frac{3x+2}{2x+3} = 4$, both sides get multiplied by $(2x + 3)$. This will cancel on the left side to yield:

$$3x + 2 = 4(2x + 3)$$

Then:

$$3x + 2 = 8x + 12$$

Now, subtract $8x$ from both sides, which yields $-5x + 2 = 12$. Subtracting 2 from both sides results in $-5x = 10$. Finally, both sides get divided by -5 to obtain $x = -2$.

Sometimes, when solving rational equations, it can be easier to try to simplify the rational expression by factoring the numerator and denominator first, then cancelling out common factors. For example, to solve $\frac{2x^2-8x+6}{x^2-3x+2} = 1$, the first step is to factor:

$$2x^2 - 8x + 6$$

$$2(x^2 - 4x + 3)$$

$$2(x - 1)(x - 3)$$

Then, factor $x^2 - 3x + 2$ into $(x - 1)(x - 2)$. This turns the original equation into:

$$\frac{2(x - 1)(x - 3)}{(x - 1)(x - 2)} = 1$$

The common factor of $(x - 1)$ can be canceled, leaving $\frac{2(x-3)}{x-2} = 1$. Now the same method used in the previous example can be followed. Multiplying both sides by $x - 2$ and performing the multiplication on the left yields $2x - 6 = x - 2$, which can be simplified to $x = 4$.

## Matrices

**Matrices** can be used to represent linear equations, solve systems of equations, and manipulate data to simulate change. Matrices consist of numerical entries in both rows and columns. The following matrix A is a $3 \times 4$ matrix because it has three rows and four columns:

$$A = \begin{bmatrix} 3 & 2 & -5 & 3 \\ 3 & 6 & 2 & -5 \\ -1 & 3 & 7 & 0 \end{bmatrix}$$

Matrices can be added or subtracted only if they have the same dimensions. For example, the following matrices can be added by adding corresponding matrix entries:

$$\begin{bmatrix} 3 & 4 \\ 2 & -6 \end{bmatrix} + \begin{bmatrix} -1 & 4 \\ 4 & 2 \end{bmatrix} = \begin{bmatrix} 2 & 8 \\ 6 & -4 \end{bmatrix}$$

Multiplication can also be used to manipulate matrices. **Scalar multiplication** involves multiplying a matrix by a constant. Each matrix entry needs to be multiplied by the constant. The following example shows a $3 \times 2$ matrix being multiplied by the constant 6:

$$6 \times \begin{bmatrix} 3 & 4 \\ 2 & -6 \\ 1 & 0 \end{bmatrix} = \begin{bmatrix} 18 & 24 \\ 12 & -36 \\ 6 & 0 \end{bmatrix}$$

Matrix multiplication of two matrices involves finding multiple dot products. The **dot product** of a row and column is the sum of the products of each corresponding row and column entry. In the following example, a $2 \times 2$ matrix is multiplied by a $2 \times 2$ matrix. The dot product of the first row and column is:

$$(2 \times 1) + (1 \times 2) = (2) + (2) = 4$$

$$\begin{bmatrix} 2 & 1 \\ 3 & 5 \end{bmatrix} \times \begin{bmatrix} 1 & 4 \\ 2 & 0 \end{bmatrix} = \begin{bmatrix} 4 & 8 \\ 13 & 12 \end{bmatrix}$$

The same process is followed to find the other three values in the solution matrix. Matrices can only be multiplied if the number of columns in the first matrix equals the number of rows in the second matrix. The previous example is also an example of square matrix multiplication because they are both square matrices. A **square matrix** has the same number of rows and columns. For square matrices, the order in which they are multiplied does matter. Therefore, matrix multiplication does not satisfy the commutative property. It does, however, satisfy the associative and distributive properties.

Another transformation of matrices can be found by using the **identity matrix**—also referred to as the **"I" matrix**. The identity matrix is similar to the number one in normal multiplication. The identity matrix is a square matrix with ones in the diagonal spots and zeros everywhere else. The identity matrix is also the result of multiplying a matrix by its inverse. This process is similar to multiplying a number by its reciprocal.

The **zero matrix** is also a matrix acting as an additive identity. The zero matrix consists of zeros in every entry. It does not change the values of a matrix when using addition.

The **inverse of a matrix** is useful for solving complex systems of equations. Not all matrices have an inverse, but this can be checked by finding the **determinant** of the matrix. If the determinant of the matrix is 0, it is not inversible. Additionally, only square matrices are inversible. To find the determinant of any matrix, each value of the first row is multiplied by the determinant of submatrix consisting of all except the row and column for that value. The results of multiplication are alternatingly subtracted and added for $3 \times 3$ or larger matrices. The determinant of a matrix can be represented with straight bars (such as $|A|$) or the function $\det(A)$, where $A$ is a matrix.

Using the **square 2 x 2 matrix**, the determinant is: $|A| = \begin{vmatrix} a & b \\ c & d \end{vmatrix} = ad - bc$

The absolute value of the determinant of matrix $A$ is equal to the area of a parallelogram with vertices $(0,0)$, $(a,b)$, $(c,d)$, and $(a+b, c+d)$.

For example, the determinant of the matrix $\begin{bmatrix} -5 & 1 \\ 3 & 4 \end{bmatrix}$ is:

$$-5(4) - 1(3) = -20 - 3 = -23$$

Using a **3 x 3 matrix** $\begin{bmatrix} a & b & c \\ d & e & f \\ g & h & i \end{bmatrix}$, the determinant is $a(ei - fh) - b(di - fg) + c(dh - eg)$.

For example, the determinant of the matrix $A = \begin{bmatrix} 2 & 0 & 1 \\ -1 & 3 & 2 \\ 2 & -2 & -1 \end{bmatrix}$ is:

$$|A| = 2(3(-1) - 2(-2)) - 0(-1(-1) - 2(2)) + 1(-1(-2) - 3(2))$$

$$|A| = 2(-3 + 4) - 0(1 - 4) + 1(2 - 6)$$

$$|A| = 2(1) - 0(-3) + 1(-4)$$

$$|A| = 2 - 0 - 4 = -2$$

The pattern continues for larger square matrices. For a matrix with real values, this can then be simplified to a real number. If the determinant is non-zero, the square matrix can be inversed.

One way to find an inverse matrix is to use the **matrix of minors**. A **minor** is the determinant of the submatrix found by excluding the row and column of that minor. The matrix formed by all the minors would be $M$. To use the previous example, the minor of the first row and column is:

$$M_a = ei - fh = 3(-1) - 2(-2) = -3 + 4 = 1$$

When dealing with larger matrices it can be inconvenient to letter the items in a matrix. Another way to refer to them is by the numbers of rows and columns in the matrix. The position of any given value in some matrix $A$ is at row $i$ and column $j$ is thus $A_{i,j}$. Using the previous example, $ie - fh$ was the minor of the first matrix item, which would be $M_{1,1} = A_{2,2}A_{3,3} - A_{2,3}A_{3,2}$. The following matrix shows all the minors:

$$M = \begin{bmatrix} 1 & -3 & -4 \\ 2 & -4 & -4 \\ -3 & 5 & 6 \end{bmatrix}$$

The next step to finding the inverse is to find the **cofactor matrix** from the matrix of minors. This is simply negating every other item in the matrix, in a checkerboard-like pattern. This is done the same for matrices of all sizes. The cofactors of $M$ are:

$$\begin{bmatrix} 1 & 3 & -4 \\ -2 & -4 & 4 \\ -3 & -5 & 6 \end{bmatrix}$$

The last steps to finding the inverse are to transpose the matrix of cofactors and divide it by the determinant of the original matrix, $|A|$. **Transposing** a matrix means turning the rows into columns and vice versa. For example, the third item of the first row would become the third item of the first column. This turns the previous cofactor matrix into an **adjoint matrix**:

$$\begin{bmatrix} 1 & -2 & -3 \\ 3 & -4 & -5 \\ -4 & 4 & 6 \end{bmatrix}$$

Dividing the transposed matrix by the determinant of our original matrix gives the inverse of matrix $A$:

$$A^{-1} = \frac{1}{|A|} \times \begin{bmatrix} 1 & -2 & -3 \\ 3 & -4 & -5 \\ -4 & 4 & 6 \end{bmatrix} = \frac{1}{-2} \times \begin{bmatrix} 1 & -2 & -3 \\ 3 & -4 & -5 \\ -4 & 4 & 6 \end{bmatrix} = \begin{bmatrix} -\frac{1}{2} & 1 & \frac{3}{2} \\ \frac{3}{2} & 2 & \frac{5}{2} \\ 2 & -2 & -3 \end{bmatrix}$$

Given a system of linear equations, a matrix can be used to represent the entire system. Operations can then be performed on the matrix to solve the system. The following system offers an example:

$$x + y + z = 4$$

$$y + 3z = -2$$

$$2x + y - 2z = 12$$

There are three variables and three equations. The coefficients in the equations can be used to form a 3 x 3 matrix:

$$\begin{bmatrix} 1 & 1 & 1 \\ 0 & 1 & 3 \\ 2 & 1 & -2 \end{bmatrix}$$

The number of rows equals the number of equations, and the number of columns equals the number of variables. The numbers on the right side of the equations can be turned into a 3 x 1 matrix. That matrix is shown here:

$$\begin{bmatrix} 4 \\ -2 \\ 12 \end{bmatrix}$$

Such a matrix can also be referred to as a **vector.** The variables are represented in a matrix of their own:

$$\begin{bmatrix} x \\ y \\ z \end{bmatrix}$$

The system can be represented by the following matrix equation:

$$\begin{bmatrix} 1 & 1 & 1 \\ 0 & 1 & 3 \\ 2 & 1 & -2 \end{bmatrix} \begin{bmatrix} x \\ y \\ z \end{bmatrix} = \begin{bmatrix} 4 \\ -2 \\ 12 \end{bmatrix}$$

Simply, this is written as $AX = B$. By using the inverse of a matrix, the solution can be found: $X = A^{-1}B$. Once the inverse of $A$ is found, it is then multiplied by $B$ to find the solution to the system: $x = 12, y = -8,$ and $z = 2$.

## Systems of Equations

A **system of equations** is a group of equations that have the same variables or unknowns. These equations can be linear, but they are not always so. Finding a solution to a system of equations means finding the values of the variables that satisfy each equation. For a linear system of two equations and two variables, there could be a single solution, no solution, or infinitely many solutions.

A single solution occurs when there is one value for $x$ and $y$ that satisfies the system. This would be shown on the graph where the lines cross at exactly one point. When there is no solution, the lines are parallel and do not ever cross. With infinitely many solutions, the equations may look different, but they are the same line. One equation will be a multiple of the other, and on the graph, they lie on top of each other.

The **process of elimination** can be used to solve a system of equations. For example, the following equations make up a system: $x + 3y = 10$ and $2x - 5y = 9$. Immediately adding these equations does not eliminate a variable, but it is possible to change the first equation by multiplying the whole equation by $-2$. This changes the first equation to $-2x - 6y = -20$. The equations can be then added to obtain $-11y = -11$. Solving for $y$ yields $y = 1$. To find the rest of the solution, 1 can be substituted in for $y$ in either original equation to find the value of $x = 7$. The solution to the system is (7, 1) because it makes both equations true, and it is the point in which the lines intersect. If the system is **dependent**—having infinitely many solutions—then both variables will cancel out when the elimination method is used, resulting in an equation that is true for many values of $x$ and $y$. Since the system is dependent, both equations can be simplified to the same equation or line.

A system can also be solved using **substitution.** This involves solving one equation for a variable and then plugging that solved equation into the other equation in the system. For example, $x - y = -2$ and $3x + 2y = 9$ can be solved using substitution. The first equation can be solved for $x$, where $x = -2 + y$. Then it can be plugged into the other equation: $3(-2 + y) + 2y = 9$. Solving for $y$ yields $-6 + 3y + 2y = 9$, where $y = 3$. If $y = 3$, then $x = 1$. This solution can be checked by plugging in these values for the variables in each equation to see if it makes a true statement.

Finally, a solution to a system of equations can be found graphically. The solution to a linear system is the point or points where the lines cross. The values of $x$ and $y$ represent the coordinates $(x, y)$ where the lines intersect. Using the same system of equations as above, they can be solved for $y$ to put them in slope-intercept form, $y = mx + b$.

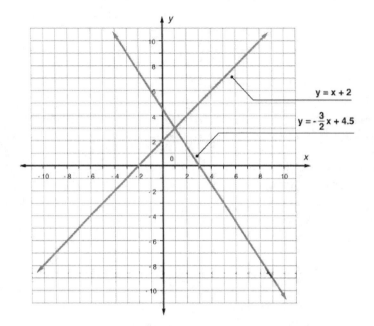

These equations become $y = x + 2$ and $y = -\frac{3}{2}x + 4.5$. This system with the solution is shown below:

A system of equations may also be made up of a linear and a quadratic equation. These systems may have one solution, two solutions, or no solutions. The graph of these systems involves one straight line and one parabola. Algebraically, these systems can be solved by solving the linear equation for one variable and plugging that answer in to the quadratic equation. If possible, the equation can then be solved to find part of the answer. The graphing method is commonly used for these types of systems. On a graph, these two lines can be found to intersect at one point, at two points across the parabola, or at no points.

Finding solutions to systems of equations is essentially finding what values of the variables make both equations true. It is finding the input value that yields the same output value in both equations. For functions $g(x)$ and $f(x)$, the equation $g(x) = f(x)$ means the output values are being set equal to each other. Solving for the value of $x$ means finding the $x$-coordinate that gives the same output in both functions. For example, $f(x) = x + 2$ and $g(x) = -3x + 10$ is a system of equations. Setting $f(x) = g(x)$ yields the equation $x + 2 = -3x + 10$. Solving for $x$, gives the $x$-coordinate $x = 2$ where the two lines cross. This value can also be found by using a table or a graph. On a table, both equations can be given the same inputs, and the outputs can be recorded to find the point(s) where the lines cross. Any method of solving finds the same solution, but some methods are more appropriate for some systems of equations than others.

Systems of **linear inequalities** are like systems of equations, but the solutions are different. Since inequalities have infinitely many solutions, their systems also have infinitely many solutions. Finding the solutions of inequalities involves graphs. A system of two equations and two inequalities is linear; thus, the lines can be graphed using slope-intercept form. If the inequality has an equals sign, the line is solid. If the inequality only has a greater than or less than symbol, the line on the graph is dotted. Dashed lines indicate that points lying on the line are not included in the solution. After the lines are graphed, a region is shaded on one side of the line. This side is found by determining if a point—known as a **test**

point—lying on one side of the line produces a true inequality. If it does, that side of the graph is shaded. If the point produces a false inequality, the line is shaded on the opposite side from the point. The graph of a system of inequalities involves shading the intersection of the two shaded regions.

# Functions

## Functions

A **function** is defined as a relationship between inputs and outputs where there is only one output value for a given input. As an example, the following function is in function notation:

$$f(x) = 3x - 4$$

The $f(x)$ represents the output value for an input of $x$. If $x = 2$, the equation becomes:

$$f(2) = 3(2) - 4$$

$$6 - 4 = 2$$

The input of 2 yields an output of 2, forming the ordered pair $(2,2)$. The following set of ordered pairs corresponds to the given function: $(2,2), (0,-4), (-2,-10)$. The set of all possible inputs of a function is its **domain**, and all possible outputs is called the **range**. By definition, each member of the domain is paired with only one member of the range.

Functions can also be defined recursively. In this form, they are not defined explicitly in terms of variables. Instead, they are defined using previously-evaluated function outputs, starting with either $f(0)$ or $f(1)$. An example of a recursively-defined function is:

$$f(1) = 2$$

$$f(n) = 2f(n-1) + 2n$$

$$n > 1$$

The domain of this function is the set of all integers.

A function $f(x)$ is a mathematical object which takes one number, $x$, as an input and gives a number in return. The input is called the **independent variable**. If the variable is set equal to the output, as in $y = f(x)$, then this is called the **dependent variable**. To indicate the dependent value a function, y, gives for a specific independent variable, x, the notation y = $f(x)$ is used.

The **domain** of a function is the set of values that the independent variable is allowed to take. Unless otherwise specified, the domain is any value for which the function is well defined. The **range** of the function is the set of possible outputs for the function.

In many cases, a function can be defined by giving an equation. For instance, $f(x) = x^2$ indicates that given a value for x, the output of f is found by squaring x.

Not all equations in x and y can be written in the form $y = f(x)$. An equation can be written in such a form if it satisfies the **vertical line test**: no vertical line meets the graph of the equation at more than a single point. In this case, y is said to be a *function of* x. If a vertical line meets the graph in two places, then this equation cannot be written in the form $y = f(x)$.

The graph of a function $f(x)$ is the graph of the equation $y = f(x)$. Thus, it is the set of all pairs $(x, y)$ where $y = f(x)$. In other words, it is all pairs $(x, f(x))$. The x-intercepts are called the **zeros** of the function. The y-intercept is given by $f(0)$.

If, for a given function $f$, the only way to get $f(a) = f(b)$ is for $a = b$, then $f$ is *one-to-one*. Often, even if a function is not one-to-one on its entire domain, it is one-to-one by considering a restricted portion of the domain.

A function $f(x) = k$ for some number $k$ is called a **constant function**. The graph of a constant function is a horizontal line.

The function $f(x) = x$ is called the **identity function**. The graph of the identity function is the diagonal line pointing to the upper right at 45 degrees, $y = x$.

A function is called **monotone** if it is either always increasing or always decreasing. For example, the functions $f(x) = 3x$ and $f(x) = -x^5$ are monotone.

An **even function** looks the same when flipped over the y-axis: $f(x) = f(-x)$. The following image shows a graphic representation of an even function.

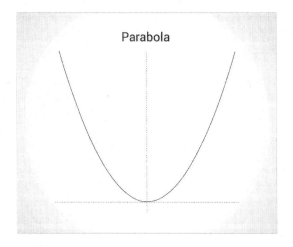

Parabola

An **odd function** looks the same when flipped over the y-axis and then flipped over the x-axis: $f(x) = -f(-x)$. The following image shows an example of an odd function.

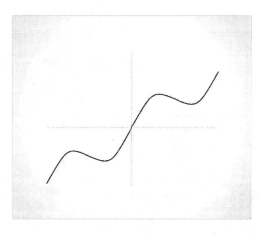

## Domain and Range

The domain and range of a function can be found visually by its plot on the coordinate plane. In the function $f(x) = x^2 - 3$, for example, the domain is all real numbers because the parabola stretches as far left and as far right as it can go, with no restrictions. This means that any input value from the real number system will yield an answer in the real number system. For the range, the inequality $y \geq -3$ would be used to describe the possible output values because the parabola has a minimum at $y = -3$. This means there will not be any real output values less than $-3$ because $-3$ is the lowest value it reaches on the y-axis.

These same answers for domain and range can be found by observing a table. The table below shows that from input values $x = -1$ to $x = 1$, the output results in a minimum of $-3$. On each side of $x = 0$, the numbers increase, showing that the range is all real numbers greater than or equal to $-3$.

| x (domain/input) | y (range/output) |
| --- | --- |
| -2 | 1 |
| -1 | -2 |
| 0 | -3 |
| -1 | -2 |
| 2 | 1 |

## Function Behavior

Different types of functions behave in different ways. A function is defined to be increasing over a subset of its domain if for all $x_1 \geq x_2$ in that interval, $f(x_1) \geq f(x_2)$. Also, a function is decreasing over an interval if for all $x_1 \geq x_2$ in that interval, $f(x_1) \leq f(x_2)$. A point in which a function changes from increasing to decreasing can also be labeled as the **maximum value** of a function if it is the largest point the graph reaches on the y-axis. A point in which a function changes from decreasing to increasing can be labeled as the minimum value of a function if it is the smallest point the graph reaches on the y-axis. Maximum values are also known as **extreme values**. The graph of a continuous function does not have any breaks or jumps in the graph. This description is not true of all functions. A radical function, for example, $f(x) = \sqrt{x}$, has a restriction for the domain and range because there are no real negative inputs or outputs for this function. The domain can be stated as $x \geq 0$, and the range is $y \geq 0$.

Logarithmic and exponential functions also have different behavior than other functions. These two types of functions are inverses of each other. The **inverse** of a function can be found by switching the place of $x$ and y, and solving for y. When this is done for the exponential equation, $y = 2^x$, the function $y = \log_2 x$ is found. The general form of a **logarithmic function** is $y = log_b x$, which says $b$ raised to the $y$ power equals $x$.

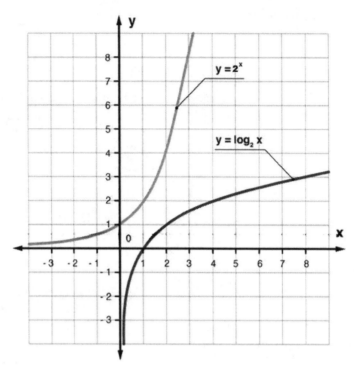

The thick black line on the graph above represents the logarithmic function $y = \log_2 x$. This curve passes through the point $(1, 0)$, just as all log functions do, because any value $b^0 = 1$. The graph of this logarithmic function starts very close to zero but does not touch the y-axis. The output value will never be zero by the definition of logarithms. The thinner gray line seen above represents the exponential function $y = 2^x$. The behavior of this function is opposite the logarithmic function because the graph of an inverse function is the graph of the original function flipped over the line $y = x$. The curve passes through the point $(0, 1)$ because any number raised to the zero power is one. This curve also gets very close to the $x$-axis but never touches it because an exponential expression never has an output of zero. The $x$-axis on this graph is called a horizontal asymptote. An **asymptote** is a line that represents a boundary for a function. It shows a value that the function will get close to, but never reach.

Functions can also be described as being even, odd, or neither. If $f(-x) = f(x)$, the function is even. For example, the function $f(x) = x^2 - 2$ is even. Plugging in $x = 2$ yields an output of $y = 2$. After changing the input to $x = -2$, the output is still $y = 2$. The output is the same for opposite inputs. Another way to observe an even function is by the symmetry of the graph. If the graph is symmetrical about the axis, then the function is even. If the graph is symmetric about the origin, then the function is odd. Algebraically, if $f(-x) = -f(x)$, the function is odd.

Also, a function can be described as **periodic** if it repeats itself in regular intervals. Common periodic functions are trigonometric functions. For example, $y = \sin x$ is a periodic function with period $2\pi$ because it repeats itself every $2\pi$ units along the $x$-axis.

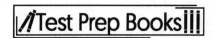

## Linear Functions

A function is called **linear** if it can take the form of the equation $f(x) = ax + b$, or $y = ax + b$, for any two numbers $a$ and $b$. A linear equation forms a straight line when graphed on the coordinate plane. An example of a linear function is shown below on the graph.

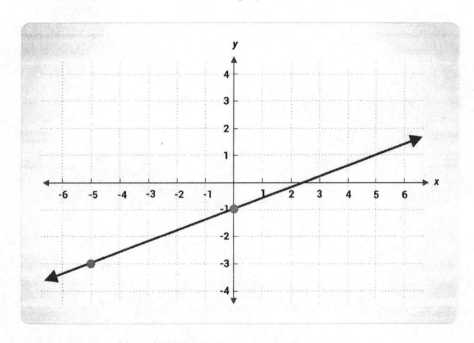

This is a graph of the following function:

$$y = \frac{2}{5}x - 1$$

A table of values that satisfies this function is shown below.

| x | y |
|---|---|
| -5 | -3 |
| 0 | -1 |
| 5 | 1 |
| 10 | 3 |

These points can be found on the graph using the form (x,y).

To graph relations and functions, the Cartesian plane is used. This means to think of the plane as being given a grid of squares, with one direction being the x-axis and the other direction the y-axis. Generally, the independent variable is placed along the horizontal axis, and the dependent variable is placed along the vertical axis. Any point on the plane can be specified by saying how far to go along the x-axis and how far along the y-axis with a pair of numbers $(x, y)$. Specific values for these pairs can be given names such as $C = (-1, 3)$. Negative values mean to move left or down; positive values mean to move right or up. The point where the axes cross one another is called the **origin.** The origin has coordinates $(0, 0)$

and is usually called *O* when given a specific label. An illustration of the Cartesian plane, along with the plotted points $(2, 1)$ and $(-1, -1)$, is below.

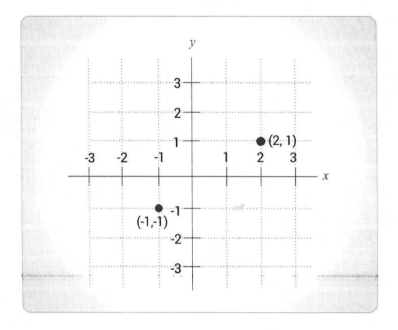

Relations also can be graphed by marking each point whose coordinates satisfy the relation. If the relation is a function, then there is only one value of *y* for any given value of *x*. This leads to the **vertical line test**: if a relation is graphed, then the relation is a function if any possible vertical line drawn anywhere along the graph would only touch the graph of the relation in no more than one place. Conversely, when graphing a function, then any possible vertical line drawn will not touch the graph of the function at any point or will touch the function at just one point. This test is made from the definition of a function, where each *x*-value must be mapped to one and only one y-value.

When graphing a linear function, note that the ratio of the change of the *y* coordinate to the change in the *x* coordinate is constant between any two points on the resulting line, no matter which two points are chosen. In other words, in a pair of points on a line, $(x_1, y_1)$ and $(x_2, y_2)$, with $x_1 \neq x_2$ so that the two points are distinct, then the ratio $\frac{y_2 - y_1}{x_2 - x_1}$ will be the same, regardless of which particular pair of points are chosen. This ratio, $\frac{y_2 - y_1}{x_2 - x_1}$, is called the *slope* of the line and is frequently denoted with the letter *m*. If slope *m* is positive, then the line goes upward when moving to the right, while if slope *m* is negative, then the line goes downward when moving to the right. If the slope is 0, then the line is called horizontal, and the y coordinate is constant along the entire line. In lines where the x coordinate is constant along the entire line, y is not actually a function of x. For such lines, the slope is not defined. These lines are called vertical lines.

Linear functions may take forms other than $y = ax + b$. The most common forms of linear equations are explained below:

1. Standard Form: $Ax + By = C$, in which the slope is given by $m = \frac{-A}{B}$, and the $y$-intercept is given by $\frac{C}{B}$.

2. Slope-Intercept Form: $y = mx + b$, where the slope is $m$ and the $y$ intercept is $b$.

3. Point-Slope Form: $y - y_1 = m(x - x_1)$, where the slope is $m$ and $(x_1, y_1)$ is any point on the chosen line.

4. Two-Point Form: $\frac{y - y_1}{x - x_1} = \frac{y_2 - y_1}{x_2 - x_1}$, where $(x_1, y_1)$ and $(x_2, y_2)$ are any two distinct points on the chosen line. Note that the slope is given by $m = \frac{y_2 - y_1}{x_2 - x_1}$.

5. Intercept Form: $\frac{x}{x_1} + \frac{y}{y_1} = 1$, in which $x_1$ is the $x$-intercept and $y_1$ is the $y$-intercept.

These five ways to write linear equations are all useful in different circumstances. Depending on the given information, it may be easier to write one of the forms over another.

If $y = mx$, $y$ is directly proportional to $x$. In this case, changing $x$ by a factor changes $y$ by that same factor. If $y = \frac{m}{x}$, $y$ is inversely proportional to $x$. For example, if $x$ is increased by a factor of 3, then $y$ will be decreased by the same factor, 3.

The **midpoint** between two points, $(x_1, y_1)$ and $(x_2, y_2)$, is given by taking the average of the $x$ coordinates and the average of the $y$ coordinates:

$$\left( \frac{x_1 + x_2}{2}, \frac{y_1 + y_2}{2} \right)$$

The **distance** between two points, $(x_1, y_1)$ and $(x_2, y_2)$, is given by the **Pythagorean formula**:

$$\sqrt{(x_2 - x_1)^2 + (y_2 - y_1)^2}$$

To find the perpendicular distance between a line $Ax + By = C$ and a point $(x_1, y_1)$ not on the line, we need to use the formula

$$\frac{|Ax_1 + By_1 + C|}{\sqrt{A^2 + B^2}}$$

## Quadratic Functions

A polynomial of degree 2 is called **quadratic.** Every quadratic function can be written in the form $ax^2 + bx + c$. The graph of a quadratic function, $y = ax^2 + bx + c$, is called a **parabola.** Parabolas are vaguely U-shaped.

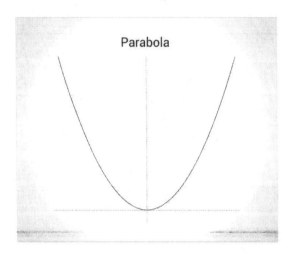

Parabola

Whether the parabola opens upward or downward depends on the sign of $a$. If $a$ is positive, then the parabola will open upward. If $a$ is negative, then the parabola will open downward. The value of $a$ will also affect how wide the parabola is. If the absolute value of $a$ is large, then the parabola will be fairly skinny. If the absolute value of $a$ is small, then the parabola will be quite wide.

Changes to the value of $b$ affect the parabola in different ways, depending on the sign of $a$. For positive values of $a$, increasing $b$ will move the parabola to the left, and decreasing $b$ will move the parabola to the right. On the other hand, if $a$ is negative, the effects will be the opposite: increasing $b$ will move the parabola to the right, while decreasing $b$ will move the parabola to the left.

Changes to the value of $c$ move the parabola vertically. The larger that $c$ is, the higher the parabola gets. This does not depend on the value of $a$.

The quantity $D = b^2 - 4ac$ is called the **discriminant** of the parabola. When the discriminant is positive, then the parabola has two real zeros, or x-intercepts. However, if the discriminant is negative, then there are no real zeros, and the parabola will not cross the x-axis. The highest or lowest point of the parabola is called the **vertex.** If the discriminant is zero, then the parabola's highest or lowest point is on the x-axis, and it will have a single real zero. The x-coordinate of the vertex can be found using the equation $x = -\frac{b}{2a}$. Plug this x-value into the equation and find the y-coordinate.

A quadratic equation is often used to model the path of an object thrown into the air. The x-value can represent the time in the air, while the y-value can represent the height of the object. In this case, the maximum height of the object would be the y-value found when the x-value is $-\frac{b}{2a}$.

Quadratic equations can be used to model real-world area problems. For example, a farmer may have a rectangular field that he needs to sow with seed. The field has length $x + 8$ and width $2x$. The formula for area should be used: $A = lw$. Therefore:

$$A = (x + 8) \times 2x = 2x^2 + 16x$$

The possible values for the length and width can be shown in a table, with input $x$ and output A. If the equation was graphed, the possible area values can be seen on the $y$-axis for given $x$-values.

## Exponential Functions

An **exponential function** is a function of the form $f(x) = b^x$, where $b$ is a positive real number other than 1. In such a function, $b$ is called the **base**.

The **domain** of an exponential function is all real numbers, and the **range** is all positive real numbers. There will always be a horizontal asymptote of $y = 0$ on one side. If $b$ is greater than 1, then the graph will be increasing when moving to the right. If $b$ is less than 1, then the graph will be decreasing when moving to the right. Exponential functions are one-to-one. The basic exponential function graph will go through the point (0, 1).

The following example demonstartes this more clearly:

Solve $5^{x+1} = 25$.

The first step is to get the $x$ out of the exponent by rewriting the equation $5^{x+1} = 5^2$ so that both sides have a base of 5. Since the bases are the same, the exponents must be equal to each other. This leaves $x + 1 = 2$ or $x = 1$. To check the answer, the $x$-value of 1 can be substituted back into the original equation.

Exponential growth and decay can be found in real-world situations. For example, if a piece of notebook paper is folded 25 times, the thickness of the paper can be found. To model this situation, a table can be used. The initial point is one-fold, which yields a thickness of 2 papers. For the second fold, the thickness is 4. Since the thickness doubles each time, the table below shows the thickness for the next few folds. Notice the thickness changes by the same factor each time. Since this change for a constant interval of folds is a factor of 2, the function is exponential. The equation for this is $y = 2^x$. For twenty-five folds, the thickness would be 33,554,432 papers.

| $x$ (folds) | $y$ (paper thickness) |
|---|---|
| 0 | 1 |
| 1 | 2 |
| 2 | 4 |
| 3 | 8 |
| 4 | 16 |
| 5 | 32 |

One exponential formula that is commonly used is the **interest formula**: $A = Pe^{rt}$. In this formula, interest is compounded continuously. $A$ is the value of the investment after the time, $t$, in years. $P$ is the initial amount of the investment, $r$ is the interest rate, and $e$ is the constant equal to approximately 2.718. Given an initial amount of \$200 and a time of 3 years, if interest is compounded continuously at a rate of 6%, the total investment value can be found by plugging each value into the formula. The invested value at the end is $239.44. In more complex problems, the final investment may be given, and the rate may be the unknown. In this case, the formula becomes $239.44 = 200e^{r3}$. Solving for $r$ requires isolating the exponential expression on one side by dividing by 200, yielding the equation $1.20 = e^{r3}$. Taking the natural log of both sides results in $\ln(1.2) = r3$. Using a calculator to evaluate the logarithmic expression, $r = 0.06 = 6\%$.

When working with logarithms and exponential expressions, it is important to remember the relationship between the two. In general, the logarithmic form is $y = \log_b x$ for an exponential form $b^y = x$. Logarithms and exponential functions are inverses of each other.

## Logarithmic Functions

A **logarithmic function** is an inverse for an exponential function. The inverse of the base $b$ exponential function is written as $\log_b(x)$, and is called the **base b logarithm**. The domain of a logarithm is all positive real numbers. It has the properties that $\log_b(b^x) = x$. For positive real values of $x$,

$$b^{\log_b(x)} = x$$

When there is no chance of confusion, the parentheses are sometimes skipped for logarithmic functions: $\log_b(x)$ may be written as $\log_b x$. For the special number $e$, the base $e$ logarithm is called the **natural logarithm** and is written as $\ln x$. Logarithms are one-to-one.

When working with logarithmic functions, it is important to remember the following properties. Each one can be derived from the definition of the logarithm as the inverse to an exponential function:

- $\log_b 1 = 0$
- $\log_b b = 1$
- $\log_b b^p = p$
- $\log_b MN = \log_b M + \log_b N$
- $\log_b \frac{M}{N} = \log_b M - \log_b N$
- $\log_b M^p = p \log_b M$

When solving equations involving exponentials and logarithms, the following fact should be used:

If $f$ is a one-to-one function, $a = b$ is equivalent to $f(a) = f(b)$.

Using this, together with the fact that logarithms and exponentials are inverses, allows for manipulations of the equations to isolate the variable as is demonstrated in the following example:

Solve $4 = \ln(x - 4)$.

Using the definition of a logarithm, the equation can be changed to $e^4 = e^{\ln(x-4)}$. The functions on the right side cancel with a result of $e^4 = x - 4$. This then gives,
$$x = 4 + e^4$$

## Rational Functions

A rational function is similar to an equation, but it includes two variables. In general, a rational function is in the form: $f(x) = \frac{p(x)}{q(x)}$, where $p(x)$ and $q(x)$ are polynomials. Rational functions are defined everywhere except where the denominator is equal to zero. When the denominator is equal to zero, this

indicates either a hole in the graph or an asymptote. An asymptote can be either vertical, horizontal, or slant. A hole occurs when both the numerator and denominator are equal to 0 for a given value of $x$. A rational function can have at most one vertical asymptote and one horizontal or slant asymptote. An asymptote is a line such that the distance between the curve and the line tends toward 0, but never reaches it, as the line heads toward infinity. Examples of these types of functions are shown below. The first graph shows a rational function with a vertical asymptote at x = 0. This can be found by setting the denominator equal to 0. In this case it is just x = 0.

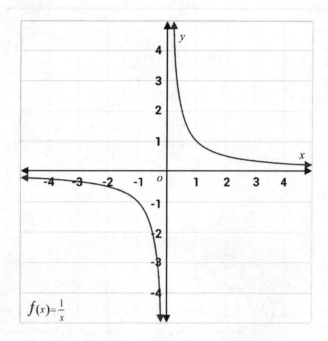

$f(x) = \frac{1}{x}$

The second graph shows a rational function with a vertical asymptote at $x = -0.5$. Again, this can be found by just setting the denominator equal to 0. So:

$$2x^2 + x = 0$$

$$2x + 1 = 0$$

$$2x = -1$$

$$x = -0.5$$

This graph also has a hole in the graph at $x = 0$. This is because both the numerator and denominator are equal to 0 when $x = 0$.

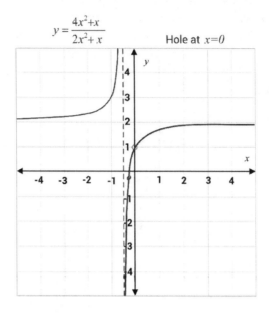

## Piecewise Functions

A piecewise-defined function also has a different appearance on the graph. In the following function, there are three equations defined over different intervals. It is a function because there is only one y-value for each $x$-value, passing the Vertical Line Test. The domain is all real numbers less than or equal to 6. The range is all real numbers greater than zero. From left to right, the graph decreases to zero, then increases to almost 4, and then jumps to 6.

From input values greater than 2, the input decreases just below 8 to 4, and then stops.

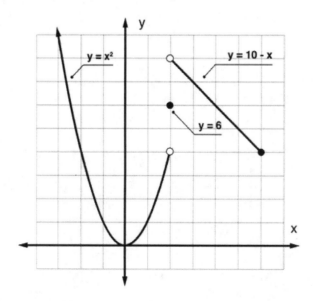

## Trigonometric Functions

**Trigonometric functions** are built out of two basic functions, the **sine** and **cosine**, written as $\sin \theta$ and $\cos \theta$, respectively. Note that similar to logarithms, it is customary to drop the parentheses as long as the result is not confusing.

Sine and cosine are defined using the **unit circle**. If $\theta$ is the angle going counterclockwise around the origin from the $x$-axis, then the point on the unit circle in that direction will have the coordinates ($\cos \theta$, $\sin \theta$).

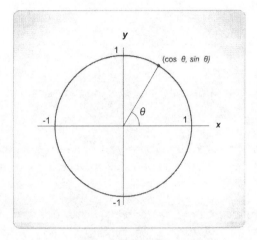

Since the angle returns to the start every $2\pi$ radians (or 360 degrees), the graph of these functions is **periodic**, with period $2\pi$. This means that the graph repeats itself as one moves along the $x$-axis because $\sin \theta = \sin(\theta + 2\pi)$. Cosine works similarly.

From the unit circle definition, the sine function starts at 0 when $\theta = 0$. It grows to 1 as $\theta$ grows to $\pi/2$, and then back to 0 at $\theta = \pi$. Then it decreases to -1 as $\theta$ grows to $3\pi/2$, and goes back up to 0 at $\theta = 2\pi$.

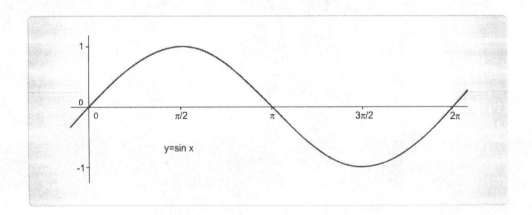

The graph of the cosine is similar. The cosine graph will start at 1, decreasing to 0 at $\pi/2$ and continuing to decrease to -1 at $\theta = \pi$. Then, it grows to 0 as $\theta$ grows to $3\pi/2$ and back up to 1 at $\theta = 2\pi$.

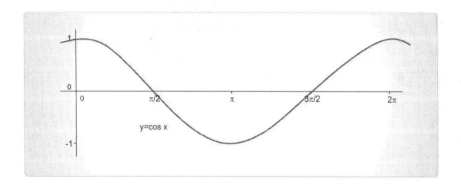

Another trigonometric function that is frequently used, is the **tangent** function. This is defined as the following equation: $\tan \theta = \frac{\sin \theta}{\cos \theta}$.

The tangent function is a period of $\pi$ rather than $2\pi$ because the sine and cosine functions have the same absolute values after a change in the angle of $\pi$, but they flip their signs. Since the tangent is a ratio of the two functions, the changes in signs cancel.

The tangent function will be zero when sine is zero, and it will have a vertical asymptote whenever cosine is zero. The following graph shows the tangent function:

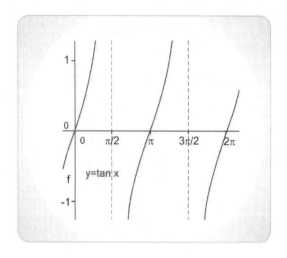

Three other trigonometric functions are sometimes useful. These are the **reciprocal** trigonometric functions, so named because they are just the reciprocals of sine, cosine, and tangent. They are the **cosecant**, defined as $\csc \theta = \frac{1}{\sin \theta}$, the **secant**, $\sec \theta = \frac{1}{\cos \theta}$, and the **cotangent**, $\cot \theta = \frac{1}{\tan \theta}$. Note that from the definition of tangent, $\cot \theta = \frac{\cos \theta}{\sin \theta}$.

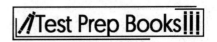

In addition, there are three identities that relate the trigonometric functions to one another:

- $\cos\theta = \sin(\frac{\pi}{2} - \theta)$
- $\csc\theta = \sec\left(\frac{\pi}{2} - \theta\right)$
- $\cot\theta = \tan(\frac{\pi}{2} - \theta)$

Here is a list of commonly-needed values for trigonometric functions, given in radians, for the first quadrant:

## Table for trigonometric functions

| | | |
|---|---|---|
| $\sin 0 = 0$ | $\cos 0 = 1$ | $\tan 0 = 0$ |
| $\sin\frac{\pi}{6} = \frac{1}{2}$ | $\cos\frac{\pi}{6} = \frac{\sqrt{3}}{2}$ | $\tan\frac{\pi}{6} = \frac{\sqrt{3}}{3}$ |
| $\sin\frac{\pi}{4} = \frac{\sqrt{2}}{2}$ | $\cos\frac{\pi}{4} = \frac{\sqrt{2}}{2}$ | $\tan\frac{\pi}{4} = 1$ |
| $\sin\frac{\pi}{3} = \frac{\sqrt{3}}{2}$ | $\cos\frac{\pi}{3} = \frac{1}{2}$ | $\tan\frac{\pi}{3} = \sqrt{3}$ |
| $\sin\frac{\pi}{2} = 1$ | $\cos\frac{\pi}{2} = 0$ | $\tan\frac{\pi}{2} = undefined$ |
| $\csc 0 = undefined$ | $\sec 0 = 1$ | $\cot 0 = undefined$ |
| $\csc\frac{\pi}{6} = 2$ | $\sec\frac{\pi}{6} = \frac{2\sqrt{3}}{3}$ | $\cot\frac{\pi}{6} = \sqrt{3}$ |
| $\csc\frac{\pi}{4} = \sqrt{2}$ | $\sec\frac{\pi}{4} = \sqrt{2}$ | $\cot\frac{\pi}{4} = 1$ |
| $\csc\frac{\pi}{3} = \frac{2\sqrt{3}}{3}$ | $\sec\frac{\pi}{3} = 2$ | $\cot\frac{\pi}{3} = \frac{\sqrt{3}}{3}$ |
| $\csc\frac{\pi}{2} = 1$ | $\sec\frac{\pi}{2} = undefined$ | $\cot\frac{\pi}{2} = 0$ |

To find the trigonometric values in other quadrants, complementary angles can be used. The **complementary angle** is the smallest angle between the $x$-axis and the given angle.

Once the complementary angle is known, the following rule is used:

For an angle $\theta$ with complementary angle $x$, the absolute value of a trigonometric function evaluated at $\theta$ is the same as the absolute value when evaluated at $x$.

The correct sign for sine and cosine is determined by the $x$ and $y$ coordinates on the unit circle.

- Sine will be positive in quadrants I and II and negative in quadrants III and IV.
- Cosine will be positive in quadrants I and IV, and negative in II and III.
- Tangent will be positive in I and III, and negative in II and IV.

The signs of the reciprocal functions will be the same as the sign of the function of which they are the reciprocal. For example:

Find $\sin\frac{3\pi}{4}$.

The reference angle must be found first. This angle is in the II quadrant, and the angle between it and the $x$-axis is $\frac{\pi}{4}$. Now, $\sin\frac{\pi}{4} = \frac{\sqrt{2}}{2}$. Since this is in the II quadrant, sine takes on positive values (the y coordinate is positive in the II quadrant). Therefore, $\sin\frac{3\pi}{4} = \frac{\sqrt{2}}{2}$.

In addition to the six trigonometric functions defined above, there are inverses for these functions. However, since the trigonometric functions are not one-to-one, one can only construct inverses for them on a restricted domain.

Usually, the domain chosen will be $[0, \pi)$ for cosine and $(-\frac{\pi}{2}, \frac{\pi}{2}]$ for sine. The inverse for tangent can use either of these domains. The inverse functions for the trigonometric functions are also called **arc functions.** In addition to being written with a -1 as the exponent to denote that the function is an inverse, they will sometimes be written with an "a" or "arc" in front of the function name, so $\cos^{-1}\theta = \mathrm{acos}\,\theta = \arccos\theta$.

When solving equations that involve trigonometric functions, there are often multiple solutions. For example, $2\sin\theta = \sqrt{2}$ can be simplified to $\sin\theta = \frac{\sqrt{2}}{2}$. This has solutions $\theta = \frac{\pi}{4}, \frac{3\pi}{4}$, but in addition, because of the periodicity, any integer multiple of $2\pi$ can also be added to these solutions to find another solution.

The full set of solutions is $\theta = \frac{\pi}{4} + 2\pi k, \frac{3\pi}{4} + 2\pi k$ for all integer values of $k$. It is very important to remember to find all possible solutions when dealing with equations that involve trigonometric functions.

The name *trigonometric* comes from the fact that these functions play an important role in the geometry of triangles, particularly right triangles. Consider the right triangle shown in this figure:

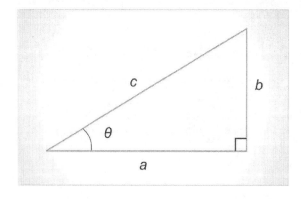

The following hold true:

- $c \sin \theta = b$
- $c \cos \theta = a$
- $\tan \theta = \dfrac{b}{a}$
- $b \csc \theta = c$
- $a \sec \theta = c$
- $\cot \theta = \dfrac{a}{b}$

It is important to remember that the angles of a triangle must add up to $\pi$ radians (180 degrees).

## Radical Functions

A radical function is any function involving a root. For instance, $y = \sqrt[n]{x}$ is a radical function with index $n$. If $n$ is odd, the function represents an odd root, and its domain and range are both all real numbers. This is because the odd root of any real number is a real number. Radical functions, for example, $f(x) = \sqrt{x}$, have a restriction for the domain and range because there are no real negative inputs or outputs for this function. The domain can be stated as $x \geq 0$, and the range is $y \geq 0$.

## Building a Function

Functions can be built out of the context of a situation. For example, the relationship between the money paid for a gym membership and the months that someone has been a member can be described through a function. If the one-time membership fee is \$40 and the monthly fee is \$30, then the function can be written $f(x) = 30x + 40$. The $x$-value represents the number of months the person has been part of the gym, while the output is the total money paid for the membership. The table below shows this relationship. It is a representation of the function because the initial cost is \$40 and the cost increases each month by \$30.

| x (months) | y (money paid to gym) |
| --- | --- |
| 0 | 40 |
| 1 | 70 |
| 2 | 100 |
| 3 | 130 |

Functions can also be built from existing functions. For example, a given function $f(x)$ can be transformed by adding a constant, multiplying by a constant, or changing the input value by a constant. The new function $g(x) = f(x) + k$ represents a vertical shift of the original function. In $f(x) = 3x - 2$, a vertical shift 4 units up would be:

$$g(x) = 3x - 2 + 4 = 3x + 2$$

Multiplying the function times a constant $k$ represents a vertical stretch, based on whether the constant is greater than or less than 1. The function

$$g(x) = kf(x) = 4(3x - 2) = 12x - 8$$

represents a stretch. Changing the input $x$ by a constant forms the function:

$$g(x) = f(x+k) = 3(x+4) - 2$$

$$3x + 12 - 2 = 3x + 10$$

and this represents a horizontal shift to the left 4 units. If $(x-4)$ was plugged into the function, it would represent a vertical shift.

A composition function can also be formed by plugging one function into another. In function notation, this is written:

$$(f \circ g)(x) = f(g(x))$$

For two functions $f(x) = x^2$ and $g(x) = x - 3$, the composition function becomes:

$$f(g(x)) = (x-3)^2$$

$$x^2 - 6x + 9$$

The composition of functions can also be used to verify if two functions are inverses of each other. Given the two functions $f(x) = 2x + 5$ and $g(x) = \frac{x-5}{2}$, the composition function can be found $(f \circ g)(x)$. Solving this equation yields:

$$f(g(x)) = 2\left(\frac{x-5}{2}\right) + 5$$

$$x - 5 + 5 = x$$

It also is true that $g(f(x)) = x$. Since the composition of these two functions gives a simplified answer of $x$, this verifies that $f(x)$ and $g(x)$ are inverse functions. The domain of $f(g(x))$ is the set of all x-values in the domain of $g(x)$ such that $g(x)$ is in the domain of $f(x)$. Basically, both $f(g(x))$ and $g(x)$ have to be defined.

To build an inverse of a function, $f(x)$ needs to be replaced with $y$, and the x and y values need to be switched. Then, the equation can be solved for y. For example, given the equation $y = e^{2x}$, the inverse can be found by rewriting the equation $x = e^{2y}$. The natural logarithm of both sides is taken down, and the exponent is brought down to form the equation:

$$\ln(x) = \ln(e)\, 2y$$

ln $(e)$=1, which yields the equation $\ln(x) = 2y$. Dividing both sides by 2 yields the inverse equation

$$\frac{\ln(x)}{2} = y = f^{-1}(x)$$

The domain of an inverse function is the range of the original function, and the range of an inverse function is the domain of the original function. Therefore, an ordered pair $(x, y)$ on either a graph or a table corresponding to $f(x)$ means that the ordered pair $(y, x)$ exists on the graph of $f^{-1}(x)$. Basically, if $f(x) = y$, then $f^{-1}(y) = x$. For a function to have an inverse, it must be one-to-one. That means it must pass the **Horizontal Line Test**, and if any horizontal line passes through the graph of the function

twice, a function is not one-to-one. The domain of a function that is not one-to-one can be restricted to an interval in which the function is one-to-one, to be able to define an inverse function.

Functions can also be formed from combinations of existing functions.

Given $f(x)$ and $g(x)$, the following can be built:

$$f + g$$

$$f - g$$

$$fg$$

$$\frac{f}{g}$$

The domains of $f + g, f - g,$ and $fg$ are the intersection of the domains of $f$ and $g$. The domain of $\frac{f}{g}$ is the same set, excluding those values that make $g(x) = 0$.

For example, if:

$$f(x) = 2x + 3$$

$$g(x) = x + 1$$

then

$$\frac{f}{g} = \frac{2x + 3}{x + 1}$$

Its domain is all real numbers except -1.

## Comparing Functions

As mentioned, three common functions used to model different relationships between quantities are linear, quadratic, and exponential functions. **Linear functions** are the simplest of the three, and the independent variable $x$ has an exponent of 1. Written in the most common form, $y = mx + b$, the coefficient of $x$ indicates how fast the function grows at a constant rate, and the $b$-value denotes the starting point. A **quadratic** function has an exponent of 2 on the independent variable $x$. Standard form for this type of function is $y = ax^2 + bx + c$, and the graph is a parabola. These type functions grow at a changing rate. An **exponential** function has an independent variable in the exponent $y = ab^x$. The graph of these types of functions is described as **growth** or **decay**, based on whether the base, $b$, is greater than or less than 1. These functions are different from quadratic functions because the base stays constant. A common base is base $e$.

The following three functions model a linear, quadratic, and exponential function respectively: $y = 2x$, $y = x^2$, and $y = 2^x$. Their graphs are shown below. The first graph, modeling the linear function, shows that the growth is constant over each interval. With a horizontal change of 1, the vertical change is 2. It models a constant positive growth. The second graph shows the quadratic function, which is a curve that is symmetric across the y-axis. The growth is not constant, but the change is mirrored over the axis. The last graph models the exponential function, where the horizontal change of 1 yields a vertical change that increases more and more. The exponential graph gets very close to the $x$-axis, but never touches it, meaning there is an asymptote there. The y-value can never be zero because the base of 2 can never be raised to an input value that yields an output of zero.

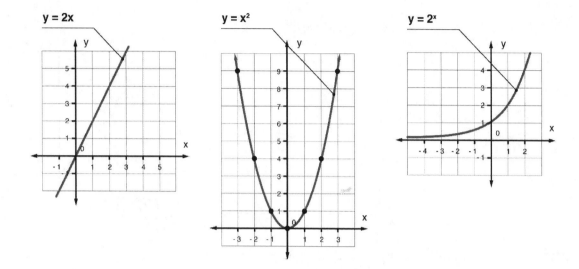

The three tables below show specific values for three types of functions. The third column in each table shows the change in the y-values for each interval. The first table shows a constant change of 2 for each equal interval, which matches the slope in the equation $y = 2x$. The second table shows an increasing change, but it also has a pattern. The increase is changing by 2 more each time, so the change is quadratic. The third table shows the change as factors of the base, 2. It shows a continuing pattern of factors of the base.

| $y = 2x$ | | |
|---|---|---|
| $x$ | $y$ | $\Delta y$ |
| 1 | 2 | |
| 2 | 4 | 2 |
| 3 | 6 | 2 |
| 4 | 8 | 2 |
| 5 | 10 | 2 |

| $y = x^2$ | | |
|---|---|---|
| $x$ | $y$ | $\Delta y$ |
| 1 | 1 | |
| 2 | 4 | 3 |
| 3 | 9 | 5 |
| 4 | 16 | 7 |
| 5 | 25 | 9 |

| $y = 2^x$ | | |
|---|---|---|
| $x$ | $y$ | $\Delta y$ |
| 1 | 2 | |
| 2 | 4 | 2 |
| 3 | 8 | 4 |
| 4 | 16 | 8 |
| 5 | 32 | 16 |

Given a table of values, the type of function can be determined by observing the change in $y$ over equal intervals. For example, the tables below model two functions. The changes in interval for the $x$-values is 1 for both tables. For the first table, the $y$-values increase by 5 for each interval. Since the change is constant, the situation can be described as a linear function. The equation would be $y = 5x + 3$. For the second table, the change for $y$ is 5, 20, 100, and 500, respectively. The increases are multiples of 5,

meaning the situation can be modeled by an exponential function. The equation $y = 5^x + 3$ models this situation.

| x | y |
|---|---|
| 0 | 3 |
| 1 | 8 |
| 2 | 13 |
| 3 | 18 |
| 4 | 23 |

| x | y |
|---|---|
| 0 | 3 |
| 1 | 8 |
| 2 | 28 |
| 3 | 128 |
| 4 | 628 |

## Evaluating Functions

To evaluate functions, plug in the given value everywhere the variable appears in the expression for the function. For example, find $g(-2)$ where $g(x) = 2x^2 - \frac{4}{x}$. To complete the problem, plug in -2 in the following way:

$$g(-2) = 2(-2)^2 - \frac{4}{-2}$$

$$2 \times 4 + 2$$

$$8 + 2 = 10$$

# *Geometry*

## Shapes and Solids

A **polygon** is a closed geometric figure in a plane (flat surface) consisting of at least 3 sides formed by line segments. These are often defined as two-dimensional shapes. Common two-dimensional shapes include circles, triangles, squares, rectangles, pentagons, and hexagons. Note that a circle is a two-dimensional shape without sides.

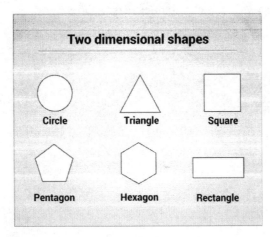

Polygons can be either convex or concave. A polygon that has interior angles all measuring less than 180° is convex. A concave polygon has one or more interior angles measuring greater than 180°. Examples are shown below.

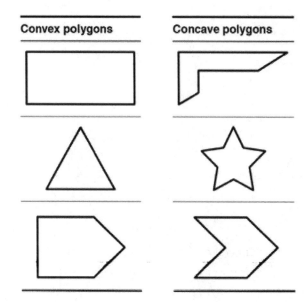

Polygons can be classified by the number of sides (also equal to the number of angles) they have. The following are the names of polygons with a given number of sides or angles:

| # of Sides | Name of Polygon |
|------------|-----------------|
| 3 | Triangle |
| 4 | Quadrilateral |
| 5 | Pentagon |
| 6 | Hexagon |
| 7 | Septagon (or heptagon) |
| 8 | Octagon |
| 9 | Nonagon |
| 10 | Decagon |

Equiangular polygons are polygons in which the measure of every interior angle is the same. The sides of equilateral polygons are always the same length. If a polygon is both equiangular and equilateral, the polygon is defined as a regular polygon.

Triangles can be further classified by their sides and angles. A triangle with its largest angle measuring 90° is a right triangle. A triangle with the largest angle less than 90° is an acute triangle. A triangle with the largest angle greater than 90° is an obtuse triangle. Below is an example of a right triangle.

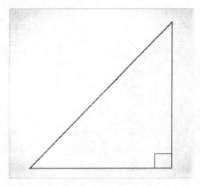

A triangle consisting of two equal sides and two equal angles is an isosceles triangle. A triangle with three equal sides and three equal angles is an equilateral triangle. A triangle with no equal sides or angles is a scalene triangle.

Isosceles triangle:

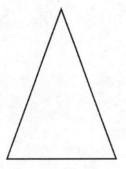

Equilateral triangle:

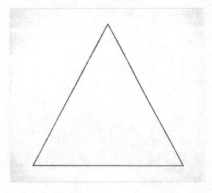

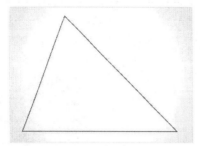

Scalene triangle:

Quadrilaterals can be further classified according to their sides and angles. A quadrilateral with exactly one pair of parallel sides is called a trapezoid. A quadrilateral that shows both pairs of opposite sides parallel is a parallelogram. Parallelograms include rhombuses, rectangles, and squares. A rhombus has four equal sides. A rectangle has four equal angles (90° each). A square has four 90° angles and four equal sides. Therefore, a square is both a rhombus and a rectangle.

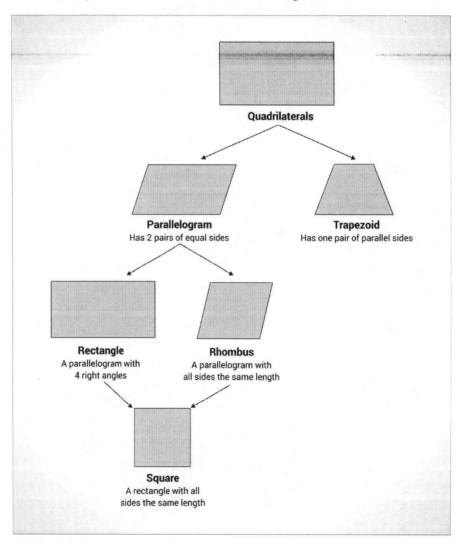

## Angles and Diagonals

**Diagonals** are lines (excluding sides) that connect two vertices within a polygon. **Mutually bisecting diagonals** intersect at their midpoints. Parallelograms, rectangles, squares, and rhombuses have mutually bisecting diagonals. However, trapezoids don't have such lines. **Perpendicular diagonals** occur when they form four right triangles at their point of intersection. Squares and rhombuses have perpendicular diagonals, but trapezoids, rectangles, and parallelograms do not. Finally, **perpendicular bisecting diagonals** (also known as **perpendicular bisectors**) form four right triangles at their point of intersection, but this intersection is also the midpoint of the two lines. Both rhombuses and squares have perpendicular bisecting angles, but trapezoids, rectangles, and parallelograms do not. Knowing these definitions can help tremendously in problems that involve both angles and diagonals.

## Polygons with More than Four Sides

A **pentagon** is a five-sided figure. A six-sided shape is a **hexagon**. A seven-sided figure is classified as a **heptagon**, and an eight-sided figure is called an **octagon**. An important characteristic is whether a polygon is regular or irregular. If it's **regular,** the side lengths and angle measurements are all equal. An **irregular** polygon has unequal side lengths and angle measurements. Mathematical problems involving polygons with more than four sides usually involve side length and angle measurements. The sum of all internal angles in a polygon equals $180(n-2)$ degrees, where $n$ is the number of sides. Therefore, the total of all internal angles in a pentagon is 540 degrees because there are five sides so $180(5-2) = 540$ degrees. Unfortunately, area formulas don't exist for polygons with more than four sides. However, their shapes can be split up into triangles, and the formula for area of a triangle can be applied and totaled to obtain the area for the entire figure.

## Solids

A solid is a three-dimensional figure that encloses a part of space. Common three-dimensional shapes include spheres, prisms, cubes, pyramids, cylinders, and cones.

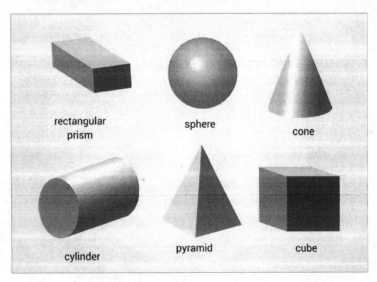

Solids consisting of all flat surfaces that are polygons are called polyhedrons. The two-dimensional surfaces that make up a polyhedron are called faces. Types of polyhedrons include prisms and pyramids. A prism consists of two parallel faces that are congruent (or the same shape and same size), and lateral faces going around (which are parallelograms). A prism is further classified by the shape of its base, as shown below:

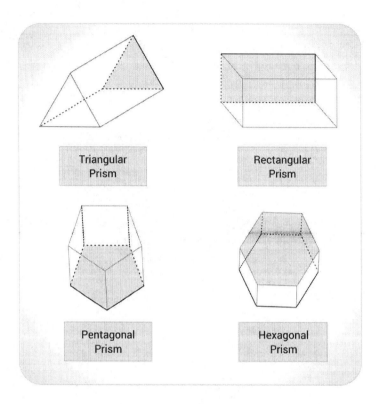

A pyramid consists of lateral faces (triangles) that meet at a common point called the vertex and one other face that is a polygon, called the base. A pyramid can be further classified by the shape of its base, as shown below.

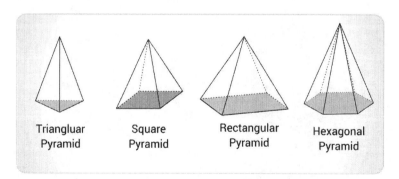

A tetrahedron is another name for a triangular pyramid. All the faces of a tetrahedron are triangles.

Solids that are not polyhedrons include spheres, cylinders, and cones. A sphere is the set of all points a given distance from a given center point. A sphere is commonly thought of as a three-dimensional circle. A cylinder consists of two parallel, congruent (same size) circles and a lateral curved surface. A cone consists of a circle as its base and a lateral curved surface that narrows to a point called the vertex.

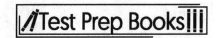

Similar polygons are the same shape but different sizes. More specifically, their corresponding angle measures are congruent (or equal) and the length of their sides is proportional. For example, all sides of one polygon may be double the length of the sides of another. Likewise, similar solids are the same shape but different sizes. Any corresponding faces or bases of similar solids are the same polygons that are proportional by a consistent value.

## Congruence and Similarity

Sometimes, two figures are similar, meaning they have the same basic shape and the same interior angles, but they have different dimensions. If the ratio of two corresponding sides is known, then that ratio, or scale factor, holds true for all of the dimensions of the new figure.

Likewise, triangles are similar if they have the same angle measurements, and their sides are proportional to one another. Triangles are **congruent** if the angles of the triangles are equal in measurement and the sides of the triangles are equal in measurement.

There are five ways to show that triangles are congruent:

1. SSS (Side-Side-Side Postulate) – when all three corresponding sides are equal in length, then the two triangles are congruent.

2. SAS (Side-Angle-Side Postulate) – if a pair of corresponding sides and the angle in between those two sides are equal, then the two triangles are congruent.

3. ASA (Angle-Side-Angle Postulate) – if a pair of corresponding angles are equal and the side lengths within those angles are equal, then the two triangles are equal.

4. AAS (Angle-Angle-Side Postulate) – when a pair of corresponding angles for two triangles and a non-included side are equal, then the two triangles are congruent.

5. HL (Hypotenuse-Leg Theorem) – if two right triangles have the same hypotenuse length, and one of the other sides in each triangle are of the same length, then the two triangles are congruent.

If two triangles are discovered to be similar or congruent, this information can assist in determining unknown parts of triangles, such as missing angles and sides.

The example below involves the question of congruent triangles. The first step is to examine whether the triangles are congruent. If the triangles are congruent, then the measure of a missing angle can be found.

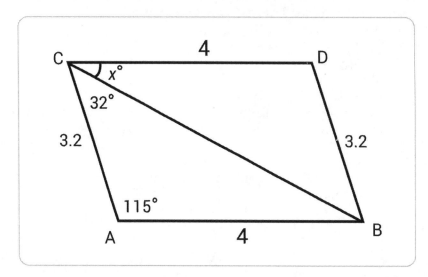

The above diagram provides values for angle measurements and side lengths in triangles *CAB* and *CDB*. Note that side *CA* is 3.2 and side *DB* is 3.2. Side *CD* is 4 and side *AB* is 4. Furthermore, line *CB* is congruent to itself by the reflexive property. Therefore, the two triangles are congruent by SSS (Side-Side-Side). Because the two triangles are congruent, all of the corresponding parts of the triangles are also congruent. Therefore, angle *x* is congruent to the inside of the angle for which a measurement is not provided in triangle *CAB*. Thus, 115º + 32º = 147º. A triangle's angles sum to 180º, therefore, 180º – 147º = 33º. Angle *x* = 33º, because the two triangles are reversed.

## Transformations of a Plane

Given a figure drawn on a plane, many changes can be made to that figure, including rotation, translation, and reflection. **Rotations** turn the figure about a point, **translations** slide the figure, and **reflections** flip the figure over a specified line. When performing these transformations, the original figure is called the **pre-image**, and the figure after transformation is called the **image**.

More specifically, **translation** means that all points in the figure are moved in the same direction by the same distance. In other words, the figure is slid in some fixed direction. Of course, while the entire figure is slid by the same distance, this does not change any of the measurements of the figures involved. The result will have the same distances and angles as the original figure.

In terms of Cartesian coordinates, a translation means a shift of each of the original points $(x, y)$ by a fixed amount in the *x* and *y* directions, to become:

$$(x + a, y + b)$$

Another procedure that can be performed is called **reflection**. To do this, a line in the plane is specified, called the **line of reflection**. Then, take each point and flip it over the line so that it is the same distance from the line but on the opposite side of it. This does not change any of the distances or angles involved, but it does reverse the order in which everything appears.

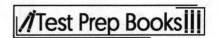

To reflect something over the x-axis, the points $(x, y)$ are sent to $(x, -y)$. To reflect something over the y-axis, the points $(x, y)$ are sent to the points $(-x, y)$. Flipping over other lines is not something easy to express in Cartesian coordinates. However, by drawing the figure and the line of reflection, the distance to the line and the original points can be used to find the reflected figure.

Example: Reflect this triangle with vertices (-1, 0), (2, 1), and (2, 0) over the y-axis. The pre-image is shown below.

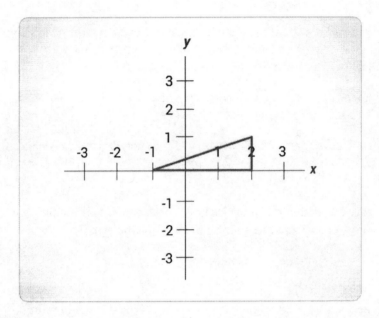

To do this, flip the x values of the points involved to the negatives of themselves, while keeping the y values the same. The image is shown here.

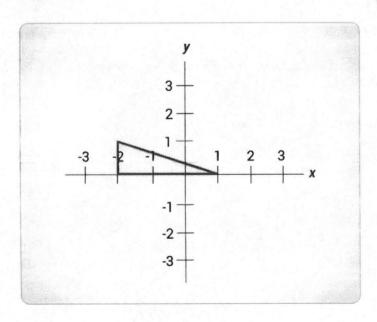

The new vertices will be (1, 0), (-2, 1), and (-2, 0).

Another procedure that does not change the distances and angles in a figure is **rotation**. In this procedure, pick a center point, then rotate every vertex along a circle around that point by the same angle. This procedure is also not easy to express in Cartesian coordinates, and this is not a requirement on this test. However, as with reflections, it's helpful to draw the figures and see what the result of the rotation would look like. This transformation can be performed using a compass and protractor.

Each one of these transformations can be performed on the coordinate plane without changes to the original dimensions or angles.

If two figures in the plane involve the same distances and angles, they are called **congruent figures**. In other words, two figures are congruent when they go from one form to another through reflection, rotation, and translation, or a combination of these.

Remember that rotation and translation will give back a new figure that is identical to the original figure, but reflection will give back a mirror image of it.

To recognize that a figure has undergone a rotation, check to see that the figure has not been changed into a mirror image, but that its orientation has changed (that is, whether the parts of the figure now form different angles with the *x* and *y* axes).

To recognize that a figure has undergone a translation, check to see that the figure has not been changed into a mirror image, and that the orientation remains the same.

To recognize that a figure has undergone a reflection, check to see that the new figure is a mirror image of the old figure.

Keep in mind that sometimes a combination of translations, reflections, and rotations may be performed on a figure.

## Dilation

A **dilation** is a transformation that preserves angles, but not distances. This can be thought of as stretching or shrinking a figure. If a dilation makes figures larger, it is called an **enlargement**. If a dilation makes figures smaller, it is called a **reduction**. The easiest example is to dilate around the origin. In this case, multiply the *x* and *y* coordinates by a **scale factor**, $k$, sending points $(x, y)$ to $(kx, ky)$.

As an example, draw a dilation of the following triangle, whose vertices will be the points (-1, 0), (1, 0), and (1, 1).

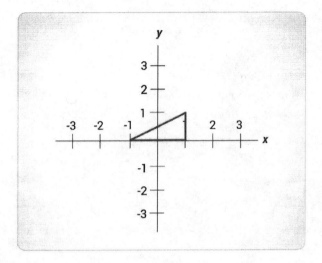

For this problem, dilate by a scale factor of 2, so the new vertices will be (-2, 0), (2, 0), and (2, 2).

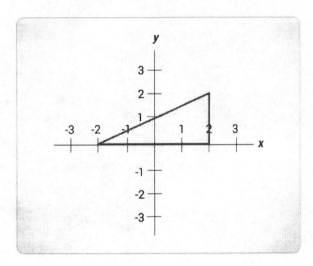

Note that after a dilation, the distances between the vertices of the figure will have changed, but the angles remain the same. The two figures that are obtained by dilation, along with possibly translation, rotation, and reflection, are all *similar* to one another. Another way to think of this is that similar figures have the same number of vertices and edges, and their angles are all the same. Similar figures have the same basic shape but are different in size.

## Surface Area and Volume

**Surface area** and volume are two- and three-dimensional measurements. Surface area measures the total surface space of an object, like the six sides of a cube. Questions about surface area will ask how much of something is needed to cover a three-dimensional object, like wrapping a present. **Volume** is the measurement of how much space an object occupies, like how much space is in the cube. Volume questions will ask how much of something is needed to completely fill the object. The most common surface area and volume questions deal with spheres, cubes, and rectangular prisms.

The formula for a cube's surface area is $SA = 6 \times s^2$, where $s$ is the length of a side. A cube has 6 equal sides, so the formula expresses the area of all the sides. Volume is simply measured by taking the cube of the length, so the formula is $V = s^3$.

The surface area formula for a rectangular prism or a general box is SA = $2(lw + lh + wh)$, where $l$ is the length, $h$ is the height, and $w$ is the width. The volume formula is $V = l \times w \times h$, which is the cube's volume formula adjusted for the unequal lengths of a box's sides.

The formula for a sphere's surface area is $SA = 4\pi r^2$, where $r$ is the sphere's radius. The surface area formula is the area for a circle multiplied by four. To measure volume, the formula is:

$$V = \frac{4}{3}\pi r^3$$

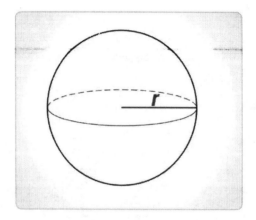

A **rectangular pyramid** is a figure with a rectangular base and four triangular sides that meet at a single vertex. If the rectangle has sides of lengths $x$ and $y$, then the volume will be given by $V = \frac{1}{3}xyh$.

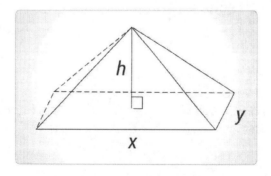

To find the surface area, the dimensions of each triangle must be known. However, these dimensions can differ depending on the problem in question. Therefore, there is no general formula for calculating total surface area.

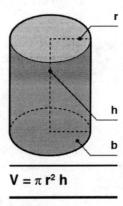

$$V = \pi r^2 h$$

The formula to find the volume of a cylinder is $\pi r^2 h$. This formula contains the formula for the area of a circle ($\pi r^2$) because the base of a cylinder is a circle. To calculate the volume of a cylinder, the slices of circles needed to build the entire height of the cylinder are added together. For example, if the radius is 5 feet and the height of the cylinder is 10 feet, the cylinder's volume is calculated by using the following equation: $\pi 5^2 \times 10$. Substituting 3.14 for $\pi$, the volume is 785.4 ft³.

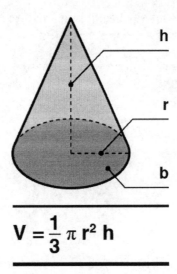

$$V = \frac{1}{3} \pi r^2 h$$

The formula used to calculate the volume of a cone is $\frac{1}{3}\pi r^2 h$. Essentially, the area of the base of the cone is multiplied by the cone's height. In a real-life example where the radius of a cone is 2 meters and the height of a cone is 5 meters, the volume of the cone is calculated by utilizing the formula:

$$\frac{1}{3}\pi 2^2 \times 5 = 21 \; m^3$$

## Solving for Missing Values in Shapes

**Perimeter** is the distance measurement around something. It can be thought of as the length of the boundary, like a fence. In contrast, area is the space occupied by a defined enclosure, like a field enclosed by a fence.

The perimeter of a square is measured by adding together all of the sides. Since a square has four equal sides, its perimeter can be calculated by multiplying the length of one side by 4. Thus, the formula is $P = 4 \times s$, where $s$ equals one side. The area of a square is calculated by squaring the length of one side, which is expressed as the formula $A = s^2$.

Like a square, a rectangle's perimeter is measured by adding together all of the sides. But as the sides are unequal, the formula is different. A rectangle has equal values for its lengths (long sides) and equal values for its widths (short sides), so the perimeter formula for a rectangle is $P = l + l + w + w = 2l + 2w$, where $l$ equals length and $w$ equals width. The area is found by multiplying the length by the width, so the formula is $A = l \times w$.

A triangle's perimeter is measured by adding together the three sides, so the formula is $P = a + b + c$, where $a, b$, and $c$ are the values of the three sides. The area is calculated by multiplying the length of the base times the height times ½, so the formula is:

$$A = \frac{1}{2} \times b \times h = \frac{bh}{2}$$

The base is the bottom of the triangle, and the height is the distance from the base to the peak. If a problem asks one to calculate the area of a triangle, it will provide the base and height.

A circle's perimeter—also known as its **circumference**—is measured by multiplying the **diameter** (the straight line measured from one side, through the center, to the direct opposite side of the circle) by $\pi$, so the formula is $\pi \times d$. This is sometimes expressed by the formula $C = 2 \times \pi \times r$, where $r$ is the **radius** of the circle. These formulas are equivalent, as the radius equals half of the diameter. The area of a circle is calculated with the formula $A = \pi \times r^2$. The test will indicate either to leave the answer with $\pi$ attached or to calculate to the nearest decimal place, which means multiplying by 3.14 for $\pi$.

The perimeter of a parallelogram is measured by adding the lengths and widths together. Thus, the formula is the same as for a rectangle:

$$P = l + l + w + w = 2l + 2w$$

However, the area formula differs from the rectangle. For a parallelogram, the area is calculated by multiplying the length by the height: $A = h \times l$

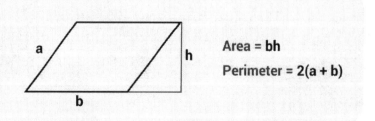

The perimeter of a trapezoid is calculated by adding the two unequal bases and two equal sides, so the formula is $P = a + b_1 + c + b_2$. Although unlikely to be a test question, the formula for the area of a trapezoid is $A = \frac{b_1 + b_2}{2} \times h$, where $h$ equals height, and $b_1$ and $b_2$ equal the bases.

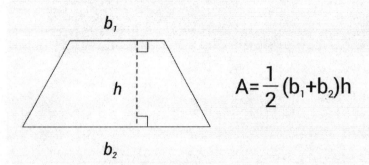

Using formulas such as perimeter and area for different shapes, it's possible to solve for missing side lengths.

Consider the following problem:

The total perimeter of a rectangular garden is 36 m. If the length of each side is 12 m, what is the width?

The formula for the perimeter of a rectangle is P=2l+2w, where P is the perimeter, l is the length, and w is the width. The first step is to substitute all of the data into the formula:

$$36 = 2(12) + 2W$$

Simplify by multiplying 2x12:

$$36 = 24 + 2W$$

Simplifying this further by subtracting 24 on each side, which gives:

$$36 - 24 = 24 - 24 + 2W$$

$$12 = 2W$$

Divide by 2:

$$6 = W$$

The width is 6 m. Remember to test this answer by substituting this value into the original formula:

$$36 = 2(12) + 2(6)$$

More complicated situations can arise where missing side lengths can be calculated by using concepts of similarity and proportional relationships. Suppose that Lara is 5 feet tall and is standing 30 feet from the base of a light pole, and her shadow is 6 feet long. How high is the light on the pole? To figure this out, it helps to make a sketch of the situation:

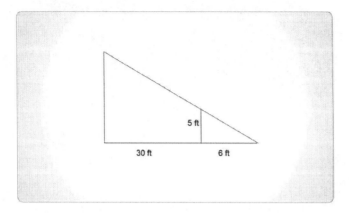

The light pole is the left side of the triangle. Lara is the 5-foot vertical line. Test takers should notice that there are two right triangles here, and that they have all the same angles as one another. Therefore, they form similar triangles. So, the ratio of proportionality between them must be found.

The bases of these triangles are known. The small triangle, formed by Lara and her shadow, has a base of 6 feet. The large triangle formed by the light pole along with the line from the base of the pole out to the end of Lara's shadow is $30 + 6 = 36$ feet long. So, the ratio of the big triangle to the little triangle is $\frac{36}{6} = 6$. The height of the little triangle is 5 feet. Therefore, the height of the big triangle will be $6 \cdot 5 = 30$ feet, meaning that the light is 30 feet up the pole.

## Composite Shapes

The perimeter of an irregular polygon is found by adding the lengths of all of the sides. In cases where all of the sides are given, this will be very straightforward, as it will simply involve finding the sum of the provided lengths. Other times, a side length may be missing and must be determined before the perimeter can be calculated.

Consider the example below:

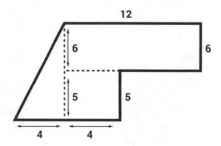

All of the side lengths are provided except for the angled side on the left. Test takers should notice that this is the hypotenuse of a right triangle. The other two sides of the triangle are provided (the base is 4 and the height is $6 + 5 = 11$). The Pythagorean Theorem can be used to find the length of the hypotenuse, remembering that $a^2 + b^2 = c^2$.

Substituting the side values provided yields:

$$(4)^2 + (11)^2 = c^2$$

Therefore, c = $\sqrt{16 + 121}$ = 11.7

Finally, the perimeter can be found by adding this new side length with the other provided lengths to get the total length around the figure:

$$4 + 4 + 5 + 8 + 6 + 12 + 11.7 = 50.7$$

Although units are not provided in this figure, remember that reporting units with a measurement is important.

The area of irregular polygons is found by decomposing, or breaking apart, the figure into smaller shapes. When the area of the smaller shapes is determined, the area of the smaller shapes will produce the area of the original figure when added together. Consider the earlier example:

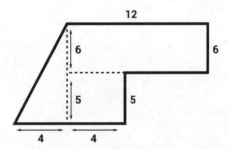

The irregular polygon is decomposed into two rectangles and a triangle. The area of the large rectangle ($A = l \times w \rightarrow A = 12 \times 6$) is 72 square units. The area of the small rectangle is 20 square units ($A = 4 \times 5$). The area of the triangle ($A = \frac{1}{2} \times b \times h \rightarrow A = \frac{1}{2} \times 4 \times 11$) is 22 square units. The sum of the areas of these figures produces the total area of the original polygon:

$$A = 72 + 20 + 22 \rightarrow A = 114 \text{ square units}$$

Here's another example:

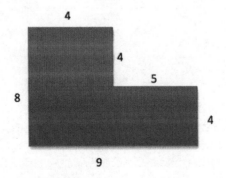

This irregular polygon is decomposed into two rectangles. The area of the large rectangle ($A = l \times w \rightarrow A = 8 \times 4$) is 32 square units. The area of the small rectangle is 20 square units ($A = 4 \times 5$). The sum of the areas of these figures produces the total area of the original polygon:

$$A = 32 + 20 \rightarrow A = 52 \text{ square units}$$

## The Pythagorean Theorem and Right Triangles

### Trigonometric Functions

From the unit circle, the trigonometric ratios were found for the special right triangle with a hypotenuse of 1.

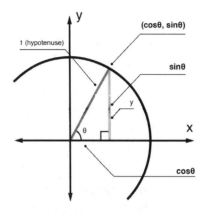

From this triangle, the following Pythagorean identities are formed:

$$\sin^2 \theta + \cos^2 \theta = 1$$

$$\tan^2 \theta + 1 = \sec^2 \theta$$

$$1 + \cot^2 \theta = \csc^2 \theta$$

The second two identities are formed by manipulating the first identity. Since identities are statements that are true for any value of the variable, then they may be used to manipulate equations. For example, a problem may ask for simplification of the expression:

$$\cos^2 x + \cos^2 x \tan^2 x$$

Using the fact that $\tan (x) = \frac{\sin x}{\cos}$, $\frac{\sin^2 x}{\cos^2 x}$ can then be substituted in for $\tan^2 x$, making the expression:

$$\cos^2 x + \cos^2 x \frac{\sin^2 x}{\cos^2 x}$$

Then the two $\cos^2 x$ terms on top and bottom cancel each other out, simplifying the expression to $\cos^2 x + \sin^2 x$. By the first Pythagorean identity stated above, the expression can be turned into $\cos^2 x + \sin^2 x = 1$.

Another set of trigonometric identities are the double-angle formulas:

$$\sin 2\alpha = 2\sin\alpha\,\cos\alpha$$

$$\cos 2\alpha = \begin{cases} \cos^2\alpha - \sin^2\alpha \\ 2\cos^2\alpha - 1 \\ 1 - 2\sin^2\alpha \end{cases}$$

Using these formulas, the following identity can be proved:

$$\sin 2x = \frac{2\tan x}{1 + \tan^2 x}$$

By using one of the Pythagorean identities, the denominator can be rewritten as:

$$1 + \tan^2 x = \sec^2 x$$

By knowing the reciprocals of the trigonometric identities, the secant term can be rewritten to form the equation:

$$\sin 2x = \frac{2\tan x}{1} * \cos^2 x$$

Replacing $\tan(x)$, the equation becomes $\sin 2x = \frac{2\sin x}{\cos x} * \cos^2 x$, where the $\cos x$ can cancel out. The new equation is:

$$\sin 2x = 2\sin x * \cos x$$

This final equation is one of the double-angle formulas.

Other trigonometric identities such as half-angle formulas, sum and difference formulas, and difference of angles formulas can be used to prove and rewrite trigonometric equations. Depending on the given equation or expression, the correct identities need to be chosen to write equivalent statements.

The graph of sine is equal to the graph of cosine, shifted $\frac{\pi}{2}$ units. Therefore, the function $y = \sin x$ is equal to:

$$y = \cos\left(\frac{\pi}{2} - x\right)$$

Within functions, adding a constant to the independent variable shifts the graph either left or right. By shifting the cosine graph, the curve lies on top of the sine function. By transforming the function, the two equations give the same output for any given input.

## Complementary Angles
Angles that add up to 90 degrees are **complementary**. Within a right triangle, two complementary angles exist because the third angle is always 90 degrees. In this scenario, the **sine** of one of the complementary angles is equal to the **cosine** of the other angle. The opposite is also true. This relationship exists because sine and cosine will be calculated as the ratios of the same side lengths.

## The Pythagorean Theorem

The **Pythagorean theorem** is an important concept in geometry. It states that for right triangles, the sum of the squares of the two shorter sides will be equal to the square of the longest side (also called the **hypotenuse**). The longest side will always be the side opposite to the 90° angle. If this side is called $c$, and the other two sides are $a$ and $b$, then the Pythagorean theorem states that:

$$c^2 = a^2 + b^2$$

Since lengths are always positive, this also can be written as:

$$c = \sqrt{a^2 + b^2}$$

A diagram to show the parts of a triangle using the Pythagorean theorem is below.

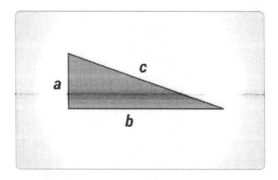

As an example of the theorem, suppose that Shirley has a rectangular field that is 5 feet wide and 12 feet long, and she wants to split it in half using a fence that goes from one corner to the opposite corner. How long will this fence need to be? To figure this out, note that this makes the field into two right triangles, whose hypotenuse will be the fence dividing it in half. Therefore, the fence length will be given by:

$$\sqrt{5^2 + 12^2} = \sqrt{169} = 13 \text{ feet long}$$

## Translating Between a Geometric Description and an Equation for a Conic Section

### Equation of a Circle

A **circle** can be defined as the set of all points that are the same distance (known as the **radius**, $r$) from a single point $C$ (known as the center of the circle). The center has coordinates $(h, k)$, and any point on the circle can be labelled with coordinates $(x, y)$.

As shown below, a **right triangle** is formed with these two points:

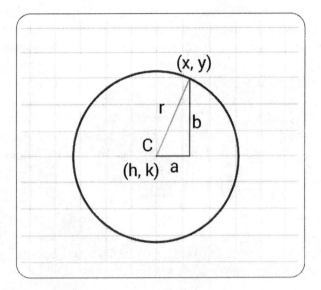

The **Pythagorean theorem** states that:

$$a^2 + b^2 = r^2$$

However, $a$ can be replaced by $|x - h|$ and $b$ can be replaced by $|y - k|$ by using the **distance formula** which is:

$$d = \sqrt{(x_2 - x_1)^2 + (y_2 - y_1)^2}$$

That substitution results in:

$$(x - h)^2 + (y - k)^2 = r^2$$

This is the formula for finding the equation of any circle with a center $(h, k)$ and a radius $r$. Note that sometimes $c$ is used instead of $r$.

## Finding the Center and Radius

Circles aren't always given in the form of the circle equation where the center and radius can be seen so easily. Oftentimes, they're given in the more general format of:

$$ax^2 + by^2 + cx + dy + e = 0$$

This can be converted to the center-radius form using the algebra technique of completing the square in both variables. First, the constant term is moved over to the other side of the equals sign, and then the $x$ and $y$ variable terms are grouped together. Then the equation is divided through by $a$ and, because this is the equation of a circle, $a = b$. At this point, the $x$-term coefficient is divided by 2, squared, and then added to both sides of the equation. This value is grouped with the $x$ terms. The same steps then need to be completed with the $y$-term coefficient. The trinomial in both $x$ and $y$ can now be factored into a square of a binomial, which gives both:

$$(x - h)^2 \text{ and}$$

$$(y - k)^2$$

99

## Parabola Equations

A **parabola** is defined as a specific type of curve such that any point on it is the same distance from a fixed point (called the **foci**) and a fixed straight line (called the **directrix**). A parabola is the shape formed from the intersection of a cone with a plane that's parallel to its side. Every parabola has an **axis of symmetry**, and its **vertex** $(h, k)$ is the point at which the axis of symmetry intersects the curve. If the parabola has an axis of symmetry parallel to the $y$-axis, the focus is the point $(h, k + f)$ and the directrix is the line $y = k - f$. For example, a parabola may have a vertex at the origin, focus $(0, f)$, and directrix $y = -f$. The equation of this parabola can be derived by using both the focus and the directrix. The distance from any coordinate on the curve to the focus is the same as the distance to the directrix, and the Pythagorean theorem can be used to find the length of $d$. The triangle has sides with length $|x|$ and $|y - f|$ and therefore:

$$d = \sqrt{x^2 + (y - f)^2}$$

By definition, the **vertex** is halfway between the focus and the directrix and $d = y + f$. Setting these two equations equal to one another, squaring each side, simplifying, and solving for $y$ gives the equation of a parabola with the focus $f$ and the vertex being the origin:

$$y = \frac{1}{4f}x^2$$

If the vertex $(h, k)$ is not the origin, a similar process can be completed to derive the equation $(x - h)^2 = 4f(y - k)$ for a parabola with focus $f$.

## Ellipse and Hyperbola Equations

An **ellipse** is the set of all points for which the sum of the distances from two fixed points (known as the *foci*) is constant. A **hyperbola** is the set of all points for which the difference between the distances from two fixed points (also known as the *foci*) is constant. The **distance formula** can be used to derive the formulas of both an ellipse and a hyperbola, given the coordinates of the foci. Consider an ellipse where its major axis is horizontal (i.e., it's longer along the $x$-axis) and its foci are the coordinates $(-c, 0)$ and $(c, 0)$. The distance from any point $(x, y)$ to $(-c, 0)$ is

$$d_1 = \sqrt{(x + c)^2 + y^2}$$

and the distance from the same point $(x, y)$ to $(c, 0)$ is:

$$d_1 = \sqrt{(x - c)^2 + y^2}$$

Using the definition of an ellipse, it's true that the sum of the distances from the vertex $a$ to each foci is equal to $d_1 + d_2$. Therefore:

$$d_1 + d_2 = (a + c) + (a - c) = 2a$$

and

$$\sqrt{(x + c)^2 + y^2} + \sqrt{(x - c)^2 + y^2} = 2a$$

After a series of algebraic steps, this equation can be simplified to $\frac{x^2}{a^2} + \frac{y^2}{b^2} = 1$, which is the equation of an ellipse with a horizontal major axis. In this case, $a > b$. When the ellipse has a vertical major axis, similar techniques result in $\frac{x^2}{b^2} + \frac{y^2}{a^2} = 1$, and $a > b$.

The equation of a hyperbola can be derived in a similar fashion. Consider a hyperbola with a horizontal major axis and its foci are also the coordinates $(-c, 0)$ and $(c, 0)$. Again, the distance from any point $(x, y)$ to $(-c, 0)$ is

$$d_1 = \sqrt{(x + c)^2 + y^2}$$

and the distance from the same point $(x, y)$ to $(c, 0)$ is:

$$d_1 = \sqrt{(x - c)^2 + y^2}$$

Using the definition of a hyperbola, it's true that the difference of the distances from the vertex $a$ to each foci is equal to $d_1 - d_2$. Therefore:

$$d_1 - d_2 = (c + a) - (c - a) = 2a$$

This means that:

$$\sqrt{(x + c)^2 + y^2} - \sqrt{(x - c)^2 + y^2} = 2a$$

After a series of algebraic steps, this equation can be simplified to:

$$\frac{x^2}{a^2} - \frac{y^2}{b^2} = 1$$

This is the equation of a hyperbola with a horizontal major axis. In this case, $a > b$. Similar techniques result in the equation $\frac{x^2}{b} - \frac{y^2}{a^2} = 1$, where $a > b$, when the hyperbola has a vertical major axis.

## Using Coordinate Geometry to Algebraically Prove Simple Geometric Theorems

<u>Proving Theorems with Coordinates</u>
Many important formulas and equations exist in geometry that use coordinates. The distance between two points $(x_1, y_1)$ and $(x_2, y_2)$ is:

$$d = \sqrt{(x_2 - x_1)^2 + (y_2 - y_1)^2}.$$

The slope of the line containing the same two points is:

$$m = \frac{y_2 - y_1}{x_2 - x_1}$$

Also, the midpoint of the line segment with endpoints $(x_1, y_1)$ and $(x_2, y_2)$ is:

$$M = \left(\frac{x_1 + x_2}{2}, \frac{y_1 + y_2}{2}\right)$$

The equations of a circle, parabola, ellipse, and hyperbola can also be used to prove theorems algebraically. Knowing when to use which formula or equation is extremely important, and knowing which formula applies to which property of a given geometric shape is an integral part of the process. In some cases, there are a number of ways to prove a theorem; however, only one way is required.

## Solving Problems with Parallel and Perpendicular Lines

Two lines can be parallel, perpendicular, or neither. If two lines are **parallel**, they have the same slope. This is proven using the idea of similar triangles. Consider the following diagram with two parallel lines, L1 and L2:

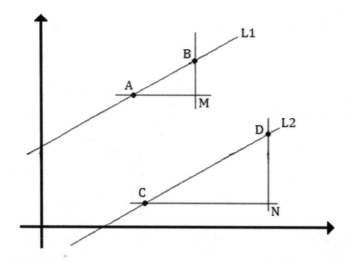

A and B are points on L1, and C and D are points on L2. Right triangles are formed with vertex M and N where lines BM and DN are parallel to the $y$-axis and AM and CN are parallel to the $x$-axis. Because all three sets of lines are parallel, the triangles are similar. Therefore:

$$\frac{BM}{DN} = \frac{MA}{NC}$$

This shows that the rise/run is equal for lines L1 and L2. Hence, their slopes are equal.

Secondly, if two lines are **perpendicular**, the product of their slopes equals -1. This means that their slopes are negative reciprocals of each other. Consider two perpendicular lines, *l* and *n*:

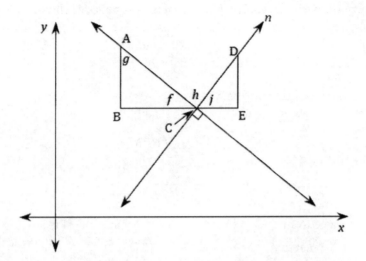

Right triangles ABC and CDE are formed so that lines BC and CE are parallel to the *x*-axis, and AB and DE are parallel to the *y*-axis. Because line BE is a straight line, angles:

$$f + h + i = 180 \ degrees$$

However, angle *h* is a right angle, so:

$$f + j = 90 \ degrees$$

By construction, $f + g = 90$, which means that $g = j$. Therefore, because angles $B = E$ and $g = j$, the triangles are similar and:

$$\frac{AB}{BC} = \frac{CE}{DE}$$

Because slope is equal to rise/run, the slope of line *l* is $-\frac{AB}{BC}$ and the slope of line *n* is $\frac{DE}{CE}$.

Multiplying the slopes together gives:

$$-\frac{AB}{BC} \cdot \frac{DE}{CE} = -\frac{CE}{DE} \cdot \frac{DE}{CE} = -1$$

This proves that the product of the slopes of two perpendicular lines equals -1. Both parallel and perpendicular lines can be integral in many geometric proofs, so knowing and understanding their properties is crucial for problem-solving.

## Formulas for Ratios

If a line segment with endpoints $(x_1, y_1)$ and $(x_2, y_2)$ is partitioned into two equal parts, the formula for **midpoint** is used. Recall this formula is:

$$M = \left( \frac{x_1 + x_2}{2}, \frac{y_1 + y_2}{2} \right)$$

The ratio of line segments is 1:1. However, if the ratio needs to be anything other than 1:1, a different formula must be used. Consider a ratio that is $a:b$. This means the desired point that partitions the line segment is $\frac{a}{a+b}$ of the way from $(x_1, y_1)$ to $(x_2, y_2)$. The actual formula for the coordinate is:

$$\left(\frac{bx_1 + ax_2}{a+b}, \frac{by_1 + ay_2}{a+b}\right)$$

## Computing Side Length, Perimeter, and Area
The side lengths of each shape can be found by plugging the endpoints into the distance formula between two ordered pairs $(x_1, y_1)$ and $(x_2, y_2)$.

As a reminder, this is the **distance formula**:

$$d = \sqrt{(x_2 - x_1)^2 + (y_2 - y_1)^2}$$

The distance formula is derived from the Pythagorean theorem. Once the side lengths are found, they can be added together to obtain the perimeter of the given polygon. Simplifications can be made for specific shapes such as squares and equilateral triangles. For example, one side length can be multiplied by 4 to obtain the perimeter of a square. Also, one side length can be multiplied by 3 to obtain the perimeter of an equilateral triangle. A similar technique can be used to calculate areas. For polygons, both side length and height can be found by using the same distance formula. Areas of triangles and quadrilaterals are straightforward through the use of $A = \frac{1}{2}bh$ or $A = bh$, depending on the shape.

To find the area of other polygons, their shapes can be partitioned into rectangles and triangles. The areas of these simpler shapes can be calculated and then added together to find the total area of the polygon.

# *Statistics and Probability*

## Center and Spread of Distributions

**Descriptive statistics** are utilized to gain an understanding of properties of a data set. This entails examining the center, spread, and shape of the sample data.

## Center
The **center** of the sample set can be represented by its mean, median or mode. The **mean** is the average of the data set. It is calculated by adding the data values together and dividing this sum by the sample size (the number of data points). The **median** is the value of the data point in the middle when the sample is arranged in numerical order. If the sample has an even number of data points, the mean of the two middle values is the median. The **mode** is the value which appears most often in a data set. It is possible to have multiple modes (if different values repeat equally as often) or no mode (if no value repeats).

## Spread

Methods for determining the **spread** of the sample include calculating the range and standard deviation for the data. The *range* is calculated by subtracting the lowest value from the highest value in the set. The **standard deviation** of the sample can be calculated using the formula:

$$\sigma = \sqrt{\frac{\sum(x - \bar{x})^2}{n - 1}}$$

$\bar{x}$ = sample mean

n = sample size

## Shape

The **shape** of the sample when displayed as a histogram or frequency distribution plot helps to determine if the sample is normally distributed (bell-shaped curve), symmetrical, or displays skewness (lack of symmetry), or kurtosis. **Kurtosis** is a measure of whether the data are heavy-tailed (high number of outliers) or light-tailed (low number of outliers).

## Data Collection Methods

**Statistical inference**, based in probability theory, makes calculated assumptions about an entire population based on data from a sample set from that population.

## Population Parameters

A population is the entire set of people or things of interest. For example, if researchers wanted to determine the number of hours of sleep per night for college females in the U.S, the population would consist of *every* college female in the country. A **sample** is a subset of the population that may be used for the study. A sample might consist of 100 students per school from 20 different colleges in the country. From the results of the survey, a sample statistic can be calculated. A **sample statistic** is a numerical characteristic of the sample data including mean and variance. A sample statistic can be used to estimate a corresponding **population parameter**, which is a numerical characteristic of the entire population.

## Confidence Intervals

A population parameter estimated using a sample statistic may be very accurate or relatively inaccurate based on errors in sampling. A **confidence interval** indicates a range of values likely to include the true population parameter. A given confidence interval such as 95% means that the true population parameter will occur within the interval for 95% of samples.

## Measurement Error

The accuracy of a population parameter based on a sample statistic may also be affected by measurement error. **Measurement error** can be divided into random error and systematic error. An example of **random error** for the previous scenario would be a student reporting 8 hours of sleep when she actually sleeps 7 hours per night. **Systematic errors** are those attributed to the measurement system. If the sleep survey gave response options of 2,4,6,8, or 10 hours. This would lead to systematic measurement error because certain values could not be accurately reported.

## Evaluating Reports and Determining the Appropriateness of Data Collection Methods

The presentation of statistics can be manipulated to produce a desired outcome. For example, in the statement "four out of five dentists recommend our toothpaste", critical readers should wonder: *who are the five dentists?* While the wording is similar, this statement is very different from "four out of every five dentists recommend our toothpaste." The context of the numerical values allows one to decipher the meaning, intent, and significance of the survey or study.

When analyzing a report, the researchers who conducted the study and their intent must be considered. Was it performed by a neutral party or by a person or group with a vested interest? The sampling method and the data collection method should also be evaluated. Was it a true random sample of the population or was one subgroup over- or underrepresented? Lastly, the measurement system used to obtain the data should be assessed. Was the system accurate and precise or was it a flawed system?

## Understanding and Modeling Relationships in Bivariate Data

In an experiment, variables are the key to analyzing data, especially when data is in a graph or table. Variables can represent anything, including objects, conditions, events, and amounts of time.

**Covariance** is a general term referring to how two variables move in relation to each other. Take for example an employee that gets paid by the hour. For them, hours worked and total pay have a positive covariance. As hours worked increases, so does pay.

**Constant variables** remain unchanged by the scientist across all trials. Because they are held constant for all groups in an experiment, they aren't being measured in the experiment, and they are usually ignored. Constants can either be controlled by the scientist directly like the nutrition, water, and sunlight given to plants, or they can be selected by the scientist specifically for an experiment like using a certain animal species or choosing to investigate only people of a certain age group.

**Independent variables** are also controlled by the scientist, but they are the same only for each group or trial in the experiment. Each group might be composed of students that all have the same color of car or each trial may be run on different soda brands. The independent variable of an experiment is what is being indirectly tested because it causes change in the dependent variables.

**Dependent variables** experience change caused by the independent variable and are what is being measured or observed. For example, college acceptance rates could be a dependent variable of an experiment that sorted a large sample of high school students by an independent variable such as test scores. In this experiment, the scientist groups the high school students by the independent variable (test scores) to see how it affects the dependent variable (their college acceptance rates).

Note that most variables can be held constant in one experiment, but also serve as the independent variable or a dependent variable in another. For example, when testing how well a fertilizer aids plant growth, its amount of sunlight should be held constant for each group of plants, but if the experiment is being done to determine the proper amount of sunlight a plant should have, the amount of sunlight is an independent variable because it is necessarily changed for each group of plants.

## Correlation

An **X-Y diagram**, also known as a scatter diagram, visually displays the relationship between two variables. The independent variable is placed on the **x-axis**, or horizontal axis, and the dependent variable is placed on the **y-axis**, or vertical axis.

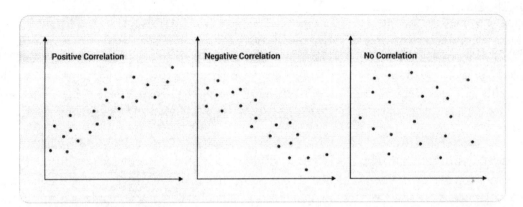

As shown in the figures above, an X-Y diagram may result in positive, negative, or no correlation between the two variables. So, in the first scatter plot as the Y factor increases the X factor increases as well. The opposite is true as well: as the X factor increases the Y factor also increases. Thus, there is a positive correlation because one factor appears to positively affect the other factor.

## Correlation Coefficient

The **correlation coefficient** (r) measures the association between two variables. Its value is between -1 and 1, where -1 represents a perfect negative linear relationship, 0 represents no relationship, and 1 represents a perfect positive linear relationship. A **negative linear relationship** means that as $x$ values increase, $y$ values decrease. A **positive linear relationship** means that as $x$ values increase, $y$ values increase. The formula for computing the correlation coefficient is:

$$r = \frac{n(\sum xy) - (\sum x)(\sum y)}{\sqrt{n(\sum x^2) - (\sum x)^2}\sqrt{n(\sum y^2) - (\Sigma y)^2}}$$

$n$ is the number of data points.

Both Microsoft Excel® and a graphing calculator can evaluate this easily once the data points are entered. A correlation greater than 0.8 or less than -0.8 is classified as "strong" while a correlation between -0.5 and 0.5 is classified as "weak."

## Calculating Probabilities, Including Related Sample Spaces

**Probability**, represented by variable *p,* always has a value from 0 to 1. The total probability for all the possible outcomes (sample space) should equal 1.

## Sample Spaces

Probabilities are based on observations of events. The probability of an event occurring is equal to the ratio of the number of favorable outcomes over the total number of possible outcomes. The total number of possible outcomes is found by constructing the sample space. The sum of probabilities of all possible distinct outcomes is equal to 1. A simple example of a sample space involves a deck of cards.

They contain 52 distinct cards, and therefore the sample space contains each individual card. To find the probability of selecting a queen on one draw from the deck, the ratio would be equal to $\frac{4}{52} = \frac{1}{13}$, which equals 4 possible queens over the total number of possibilities in the sample space.

## Verifying Independent Events

Two events aren't always independent. For examples, females with glasses and brown hair aren't independent characteristics. There definitely can be overlap because females with brown hair can wear glasses. Also, two events that exist at the same time don't have to have a relationship. For example, even if all females in a given sample are wearing glasses, the characteristics aren't related. In this case, the probability of a brunette wearing glasses is equal to the probability of a female being a brunette multiplied by the probability of a female wearing glasses. This mathematical test of $P(A \cap B) = P(A)P(B)$ verifies that two events are independent.

## Simple and Compound Events

A **simple event** consists of only one outcome. The most popular simple event is flipping a coin, which results in either heads or tails. A **compound event** results in more than one outcome and consists of more than one simple event. An example of a compound event is flipping a coin while tossing a die. The result is either heads or tails on the coin and a number from one to six on the die. The probability of a simple event is calculated by dividing the number of possible outcomes by the total number of outcomes. Therefore, the probability of obtaining heads on a coin is $^1/_2$, and the probability of rolling a 6 on a die is $^1/_6$. The probability of compound events is calculated using the basic idea of the probability of simple events. If the two events are independent, the probability of one outcome is equal to the product of the probabilities of each simple event. For example, the probability of obtaining heads on a coin and rolling a 6 is equal to:

$$^1/_2 \times {}^1/_6 = {}^1/_{12}$$

The probability of either A or B occurring is equal to the sum of the probabilities minus the probability that both A and B will occur. Therefore, the probability of obtaining either heads on a coin or rolling a 6 on a die is

$$^1/_2 + {}^1/_6 - {}^1/_{12} = {}^7/_{12}$$

The two events aren't mutually exclusive because they can happen at the same time. If two events are mutually exclusive, and the probability of both events occurring at the same time is zero, the probability of event A or B occurring equals the sum of both probabilities. An example of calculating the probability of two mutually exclusive events is determining the probability of pulling a king or a queen from a deck of cards. The two events cannot occur at the same time.

# *Integrating Essential Skills*

## Using Ratios, Rates, Proportions, and Scale Drawings to Solve Single- and Multistep Problems

Ratios, rates, proportions, and scale drawings are used when comparing two quantities. Questions on this material will include expressing relationships in simplest terms and solving for missing quantities.

## Ratios

A ratio is a comparison of two quantities that represent separate groups. For example, if a recipe calls for 2 eggs for every 3 cups of milk, it can be expressed as a ratio. Ratios can be written three ways: (1) with the word "to"; (2) using a colon; or (3) as a fraction. For the previous example, the ratio of eggs to cups of milk can be written as: 2 to 3, 2:3, or $\frac{2}{3}$. When writing ratios, the order is important. The ratio of eggs to cups of milk is not the same as the ratio of cups of milk to eggs, 3:2.

In simplest form, both quantities of a ratio should be written as integers. These should also be reduced just as a fraction would be. For example, 5:10 would reduce to 1:2. Given a ratio where one or both quantities are expressed as a decimal or fraction, both should be multiplied by the same number to produce integers. To write the ratio $\frac{1}{3}$ to 2 in simplest form, both quantities should be multiplied by 3. The resulting ratio is 1 to 6.

When a problem involving ratios gives a comparison between two groups, then: (1) a total should be provided and a part should be requested; or (2) a part should be provided and a total should be requested. Consider the following:

> The ratio of boys to girls in the 11th grade is 5:4. If there is a total of 270 11th grade students, how many are girls?

To solve this, the total number of "ratio pieces" first needs to be determined. The total number of 11th grade students is divided into 9 pieces. The ratio of boys to total students is 5:9; and the ratio of girls to total students is 4:9. Knowing the total number of students, the number of girls can be determined by setting up a proportion:

$$\frac{4}{9} = \frac{x}{270}$$

Solving the proportion, it shows that there are 120 11th grade girls.

### Rates

A rate is a ratio comparing two quantities expressed in different units. A unit rate is one in which the second is one unit. Rates often include the word *per*. Examples include miles per hour, beats per minute, and price per pound. The word *per* can be represented with a / symbol or abbreviated with the letter "p" and the units abbreviated. For example, miles per hour would be written mi/h. Given a rate that is not in simplest form (second quantity is not one unit), both quantities should be divided by the value of the second quantity. Suppose a patient had 99 heartbeats in 1½ minutes. To determine the heart rate, 1½ should divide both quantities. The result is 66 bpm.

### Scale Drawings

Scale drawings are used in designs to model the actual measurements of a real-world object. For example, the blueprint of a house might indicate that it is drawn at a scale of 3 inches to 8 feet. Given one value and asked to determine the width of the house, a proportion should be set up to solve the problem. Given the scale of 3in:8ft and a blueprint width of 1 ft (12 in.), to find the actual width of the building, the proportion $\frac{3}{8} = \frac{12}{x}$ should be used. This results in an actual width of 32 ft.

## Proportions

A proportion is a statement consisting of two equal ratios. Proportions will typically give three of four quantities and require solving for the missing value. The key to solving proportions is to set them up properly. Here's a sample problem:

If 7 gallons of gas costs $14.70, how many gallons can you get for $20?

The information should be written as equal ratios with a variable representing the missing quantity:

$$\left(\frac{gallons}{cost} = \frac{gallons}{cost}\right) : \frac{7}{14.70} = \frac{x}{20}$$

To solve, cross multiply (multiply the numerator of the first ratio by the denominator of the second and vice versa) is used, and the products are set equal to each other. Cross-multiplying results in:

$$(7)(20) = (14.7)(x)$$

Solving the equation for $x$, it can be determined that 9.5 gallons of gas can be purchased for $20.

For direct proportions, as one quantity increases, the other quantity also increases. For indirect proportions (also referred to as indirect variations, inverse proportions, or inverse variations), as one quantity increases, the other decreases. Direct proportions can be written:

$$\frac{y_1}{x_1} = \frac{y_2}{x_2}$$

Conversely, indirect proportions are written:

$$y_1 x_1 = y_2 x_2$$

Here's a sample problem:

It takes 3 carpenters 10 days to build the frame of a house. How long should it take 5 carpenters to build the same frame?

In this scenario, as one quantity increases (number of carpenters), the other decreases (number of days building); therefore, this is an inverse proportion. To solve, the products of the two variables (in this scenario, the total work performed) are set equal to each other ($y_1 x_1 = y_2 x_2$). Using $y$ to represent carpenters and $x$ to represent days, the resulting equation is:

$$(3)(10) = (5)(x_2)$$

Solving for $x_2$, it is determined that it should take 5 carpenters 6 days to build the frame of the house.

## Solving Single- and Multistep Problems Involving Percentages

The word percent means "per hundred." When dealing with percentages, it may be helpful to think of the number as a value in hundredths. For example, 15% can be expressed as "fifteen hundredths" and written as $\frac{15}{100}$ or .15.

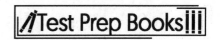

## Percent Problems

Material on percentages can include questions such as: What is 15% of 25? What percent of 45 is 3? Five is $\frac{1}{2}$% of what number? To solve these problems, the information should be rewritten as an equation where the following helpful steps are completed: (1) "what" is represented by a variable ($x$); (2) "is" is represented by an = sign; and (3) "of" is represented by multiplication. Any values expressed as a percent should be written as a decimal; and if the question is asking for a percent, the answer should be converted accordingly. Here are three sample problems based on the information above:

What is 15% of 25?

$$x = .15 \times 25$$
$$x = 3.75$$

What percent of 45 is 3?

$$x \times 45 = 3$$
$$x = 0.0\overline{6}$$
$$x = 6.\overline{6}\%$$

Five is $\frac{1}{2}$% of what number?

$$5 = .005 \times x$$
$$x = 1,000$$

## Percent Increase/Decrease

Problems dealing with percentages may involve an original value, a change in that value, and a percentage change. A problem will provide two pieces of information and ask to find the third. To do so, this formula is used:

$$\frac{change}{original\ value} \times 100 = percent\ change$$

Here's a sample problem:

Attendance at a baseball stadium has dropped 16% from last year. Last year's average attendance was 40,000. What is this year's average attendance?

Using the formula and information, the change is unknown ($x$), the original value is 40,000, and the percent change is 16%. The formula can be written as:

$$\frac{x}{40,000} \times 100 = 16$$

When solving for $x$, it is determined the change was 6,400. The problem asked for this year's average attendance, so to calculate, the change (6,400) is subtracted from last year's attendance (40,000) to determine this year's average attendance is 33,600.

## Percent More Than/Less Than

Percentage problems may give a value and what percent that given value is more than or less than an original unknown value. Here's a sample problem:

A store advertises that all its merchandise has been reduced by 25%. The new price of a pair of shoes is $60. What was the original price?

This problem can be solved by writing a proportion. Two ratios should be written comparing the cost and the percent of the original cost. The new cost is 75% of the original cost (100% - 25%); and the original cost is 100% of the original cost. The unknown original cost can be represented by $x$. The proportion would be set up as: $\frac{60}{75} = \frac{x}{100}$. Solving the proportion, it is determined the original cost was $80.

### Solving Single- and Multistep Problems Involving Measurement Quantities, Units, and Unit Conversion

Unit conversions apply to many real-world scenarios, including cooking, measurement, construction, and currency. Problems on this material can be solved similarly to those involving unit rates. Given the conversion rate, it can be written as a fraction (ratio) and multiplied by a quantity in one unit to convert it to the corresponding unit. For example, someone might want to know how many minutes are in 3½ hours. The conversion rate of 60 minutes to 1 hour can be written as:

$$\frac{60\ min}{1\ h}$$

Multiplying the quantity by the conversion rate results in:

$$3\frac{1}{2}h \times \frac{60\ min}{1\ h} = 210\ min$$

The "h" unit is canceled. To convert a quantity in minutes to hours, the fraction for the conversion rate would be flipped (to cancel the "min" unit). To convert 195 minutes to hours, the equation $195\ min \times \frac{1h}{60min}$ would be used. The result is $\frac{195h}{60}$, which reduces to $3\frac{1}{4}$ hours.

Converting units may require more than one multiplication. The key is to set up the conversion rates so that units cancel out each other and the desired unit is left. Suppose someone wants to convert 3.25 yards to inches, given that 1yd = 3ft and 12in = 1ft. To calculate, the equation use:

$$3.25yd \times \frac{3ft}{1yd} \times \frac{12in}{1ft}$$

The "yd" and "ft" units will cancel, resulting in 117 inches.

### Area, Surface Area, and Volume

The area of a two-dimensional figure refers to the number of square units needed to cover the interior region of the figure. This concept is similar to wallpaper covering the flat surface of a wall. For example, if a rectangle has an area of 21 square centimeters (written $21cm^2$), it will take 21 squares, each with sides one centimeter in length, to cover the interior region of the rectangle. Note that area is measured in square units such as: square feet or $ft^2$; square yards or $yd^2$; square miles or $mi^2$.

The surface area of a three-dimensional figure refers to the number of square units needed to cover the entire surface of the figure. This concept is similar to using wrapping paper to completely cover the outside of a box. For example, if a triangular pyramid has a surface area of 17 square inches (written $17in^2$), it will take 17 squares, each with sides one inch in length, to cover the entire surface of the pyramid. Surface area is also measured in square units.

Many three-dimensional figures (solid figures) can be represented by nets consisting of rectangles and triangles. The surface area of such solids can be determined by adding the areas of each of its faces and bases. Finding the surface area using this method requires calculating the areas of rectangles and

triangles. To find the area ($A$) of a rectangle, the length ($l$) is multiplied by the width ($w$) → $A = l \times w$. The area of a rectangle with a length of 8cm and a width of 4cm is calculated:

$$A = (8cm) \times (4cm) \rightarrow A = 32cm^2.$$

To calculate the area ($A$) of a triangle, the product of $\frac{1}{2}$, the base ($b$), and the height ($h$) is found:

$$A = \frac{1}{2} \times b \times h$$

Note that the height of a triangle is measured from the base to the vertex opposite of it forming a right angle with the base. The area of a triangle with a base of 11cm and a height of 6cm is calculated:

$$A = \frac{1}{2} \times (11cm) \times (6cm) \rightarrow A = 33cm^2$$

Consider the following triangular prism, which is represented by a net consisting of two triangles and three rectangles.

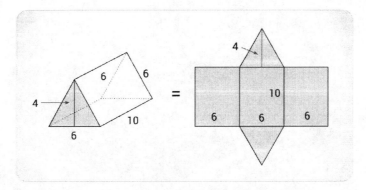

The surface area of the prism can be determined by adding the areas of each of its faces and bases. The surface area ($SA$) = area of triangle + area of triangle + area of rectangle + area of rectangle + area of rectangle.

$$SA = \left(\frac{1}{2} \times b \times h\right) + \left(\frac{1}{2} \times b \times h\right) + (l \times w) + (l \times w) + (l \times w)$$

$$SA = \left(\frac{1}{2} \times 6 \times 4\right) + \left(\frac{1}{2} \times 6 \times 4\right) + (6 \times 10) + (6 \times 10) + (6 \times 10)$$

$$SA = (12) + (12) + (60) + (60) + (60)$$

$$SA = 204 \; square \; units$$

## Effects of Changes to Dimensions on Area and Volume

Similar polygons are figures that are the same shape but different sizes. Likewise, similar solids are different sizes but are the same shape. In both cases, corresponding angles in the same positions for both figures are congruent (equal), and corresponding sides are proportional in length. For example, the triangles below are similar. The following pairs of corresponding angles are congruent: $\angle A$ and $\angle D$; $\angle B$ and $\angle E$; $\angle C$ and $\angle F$. The corresponding sides are proportional:

$$\frac{AB}{DE} = \frac{6}{3} = 2$$

$$\frac{BC}{EF} = \frac{9}{4.5} = 2$$

$$\frac{CA}{FD} = \frac{10}{5} = 2$$

In other words, triangle $ABC$ is the same shape but twice as large as triangle $DEF$.

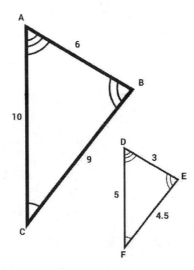

An example of similar triangular pyramids is shown below.

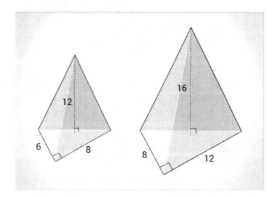

Given the nature of two- and three-dimensional measurements, changing dimensions by a given scale (multiplier) does not change the area of volume by the same scale. Consider a rectangle with a length of 5 centimeters and a width of 4 centimeters. The area of the rectangle is $20cm^2$. Doubling the dimensions of the rectangle (multiplying by a scale factor of 2) to 10 centimeters and 8 centimeters *does not* double the area to $40cm^2$. Area is a two-dimensional measurement (measured in square units).

Therefore, the dimensions are multiplied by a scale that is squared (raised to the second power) to determine the scale of the corresponding areas. For the previous example, the length and width are multiplied by 2. Therefore, the area is multiplied by $2^2$, or 4. The area of a 5cm × 4cm rectangle is $20cm^2$. The area of a 10cm × 8cm rectangle is $80cm^2$.

Volume is a three-dimensional measurement, which is measured in cubic units. Therefore, the scale between dimensions of similar solids is cubed (raised to the third power) to determine the scale between their volumes. Consider similar right rectangular prisms: one with a length of 8 inches, a width of 24 inches, and a height of 16 inches; the second with a length of 4 inches, a width of 12 inches, and a height of 8 inches. The first prism, multiplied by a scalar of $\frac{1}{2}$, produces the measurement of the second prism. The volume of the first prism, multiplied by $(\frac{1}{2})^3$, which equals $\frac{1}{8}$, produces the volume of the second prism. The volume of the first prism is 8in × 24in × 16in which equals $3,072in^3$. The volume of the second prism is $4in \times 12in \times 8in$ which equals $384in^3$:

$$3,072in^3 \times \frac{1}{8} = 384in^3$$

The rules for squaring the scalar for area and cubing the scalar for volume only hold true for similar figures. In other words, if only one dimension is changed (changing the width of a rectangle but not the length) or dimensions are changed at different rates (the length of a prism is doubled and its height is tripled) the figures are not similar (same shape). Therefore, the rules above do not apply.

## Average, Median, and Measures of Central Tendency

Suppose that $X$ is a set of data points $(x_1, x_2, x_3, \dots x_n)$ and some description of the general properties of this data need to be found.

The first property that can be defined for this set of data is the **mean**. To find the mean, add up all the data points, then divide by the total number of data points. This can be expressed using **summation notation** as:

$$\bar{X} = \frac{x_1 + x_2 + x_3 + \cdots + x_n}{n} = \frac{1}{n}\sum_{i=1}^{n} x_i$$

For example, suppose that in a class of 10 students, the scores on a test were 50, 60, 65, 65, 75, 80, 85, 85, 90, 100. Therefore, the average test score will be:

$$\frac{1}{10}(50 + 60 + 65 + 65 + 75 + 80 + 85 + 85 + 90 + 100) = 75.5$$

The mean is a useful number if the distribution of data is normal (more on this later), which roughly means that the frequency of different outcomes has a single peak and is roughly equally distributed on both sides of that peak. However, it is less useful in some cases where the data might be split or where there are some **outliers**. Outliers are data points that are far from the rest of the data. For example, suppose there are 10 executives and 90 employees at a company. The executives make $1000 per hour, and the employees make $10 per hour.

Therefore, the average pay rate will be:

$$\frac{\$1000 \times 10 + \$10 \times 90}{100} = \$109 \text{ per hour}$$

In this case, this average is not very descriptive since it's not close to the actual pay of the executives or the employees.

Another useful measurement is the **median**. In a data set $X$ consisting of data points $x_1, x_2, x_3, \ldots x_n$, the median is the point in the middle. The middle refers to the point where half the data comes before it and half comes after, when the data is recorded in numerical order. If $n$ is odd, then the median is:

$$x_{\frac{n+1}{2}}$$

If $n$ is even, it is defined as:

$$\frac{1}{2}\left(x_{\frac{n}{2}} + x_{\frac{n}{2}+1}\right)$$

It is the mean of the two data points closest to the middle of the data points. In the previous example of test scores, the two middle points are 75 and 80. Since there is no single point, the average of these two scores needs to be found. The average is:

$$\frac{75 + 80}{2} = 77.5$$

The median is generally a good value to use if there are a few outliers in the data. It prevents those outliers from affecting the "middle" value as much as when using the mean.

Since an outlier is a data point that is far from most of the other data points in a data set, this means an outlier also is any point that is far from the median of the data set. The outliers can have a substantial effect on the mean of a data set, but they usually do not change the median or mode, or do not change them by a large quantity. For example, consider the data set (3, 5, 6, 6, 6, 8). This has a median of 6 and a mode of 6, with a mean of $\frac{34}{6} \approx 5.67$. Now, suppose a new data point of 1000 is added so that the data set is now (3, 5, 6, 6, 6, 8, 1000). The median and mode, which are both still 6, remain unchanged. However, the average is now $\frac{1034}{7}$, which is approximately 147.7. In this case, the median and mode will be better descriptions for most of the data points.

The reason for outliers in a given data set is a complicated problem. It is sometimes the result of an error by the experimenter, but often they are perfectly valid data points that must be taken into consideration.

One additional measure to define for $X$ is the **mode**. This is the data point that appears more frequently. If two or more data points all tie for the most frequent appearance, then each of them is considered a mode. In the case of the test scores, where the numbers were 50, 60, 65, 65, 75, 80, 85, 85, 90, 100, there are two modes: 65 and 85.

The **first quartile** of a set of data $X$ refers to the largest value from the first ¼ of the data points. In practice, there are sometimes slightly different definitions that can be used, such as the median of the

first half of the data points (excluding the median itself if there are an odd number of data points). The term also has a slightly different use: when it is said that a data point lies in the first quartile, it means it is less than or equal to the median of the first half of the data points. Conversely, if it lies *at* the first quartile, then it is equal to the first quartile.

When it is said that a data point lies in the **second quartile**, it means it is between the first quartile and the median.

The **third quartile** refers to data that lies between ½ and ¾ of the way through the data set. Again, there are various methods for defining this precisely, but the simplest way is to include all of the data that lie between the median and the median of the top half of the data.

Data that lies in the **fourth quartile** refers to all of the data above the third quartile.

**Percentiles** may be defined in a similar manner to quartiles. Generally, this is defined in the following manner:

If a data point lies *in* the n-th percentile, this means it lies in the range of the first *n*% of the data.

If a data point lies *at* the *n*-th percentile, then it means that *n*% of the data lies below this data point.

Given a data set *X* consisting of data points $(x_1, x_2, x_3, \dots x_n)$, the **variance of X** is defined to be:

$$\frac{\sum_{i=1}^{n}(x_i - \bar{X})^2}{n}$$

This means that the variance of *X* is the average of the squares of the differences between each data point and the mean of *X*. In the formula, $\bar{X}$ is the mean of the values in the data set, and $x_i$ represents each individual value in the data set. The sigma notation indicates that the sum should be found with $n$ being the number of values to add together. $i = 1$ means that the values should begin with the first value.

Given a data set *X* consisting of data points $(x_1, x_2, x_3, \dots x_n)$, the **standard deviation of X** is defined to be

$$s_x = \sqrt{\frac{\sum_{i=1}^{n}(x_i - \bar{X})^2}{n}}$$

In other words, the standard deviation is the square root of the variance.

Both the variance and the standard deviation are measures of how much the data tend to be spread out. When the standard deviation is low, the data points are mostly clustered around the mean. When the standard deviation is high, this generally indicates that the data are quite spread out, or else that there are a few substantial outliers.

As a simple example, compute the standard deviation for the data set (1, 3, 3, 5). First, compute the mean, which will be:

$$\frac{1 + 3 + 3 + 5}{4} = \frac{12}{4} = 3$$

Now, find the variance of $X$ with the formula:

$$\sum_{i=1}^{4}(x_i - \bar{X})^2 = (1-3)^2 + (3-3)^2 + (3-3)^2 + (5-3)^2$$

$$-2^2 + 0^2 + 0^2 + 2^2 = 8$$

Therefore, the variance is $\frac{8}{4} = 2$. Taking the square root, the standard deviation will be $\sqrt{2}$.

Note that the standard deviation only depends upon the mean, not upon the median or mode(s). Generally, if there are multiple modes that are far apart from one another, the standard deviation will be high. A high standard deviation does not always mean there are multiple modes, however.

## Representing Numbers in Various Ways

### Concrete Models

Concrete objects are used to develop a tangible understanding of operations of rational numbers. Tools such as tiles, blocks, beads, and hundred charts are used to model problems. For example, a hundred chart ($10 \times 10$) and beads can be used to model multiplication. If multiplying 5 by 4, beads are placed across 5 rows and down 4 columns producing a product of 20. Similarly, tiles can be used to model division by splitting the total into equal groups. If dividing 12 by 4, 12 tiles are placed one at a time into 4 groups. The result is 4 groups of 3. This is also an effective method for visualizing the concept of remainders.

Representations of objects can be used to expand on the concrete models of operations. Pictures, dots, and tallies can help model these concepts. Utilizing concrete models and representations creates a foundation upon which to build an abstract understanding of the operations.

### Rational Numbers on a Number Line

A number line typically consists of integers (...3, 2, 1, 0, -1, -2, -3...), and is used to visually represent the value of a rational number. Each rational number has a distinct position on the line determined by comparing its value with the displayed values on the line. For example, if plotting -1.5 on the number line below, it is necessary to recognize that the value of -1.5 is .5 less than -1 and .5 greater than -2. Therefore, -1.5 is plotted halfway between -1 and -2.

Number lines can also be useful for visualizing sums and differences of rational numbers. Adding a value indicates moving to the right (values increase to the right), and subtracting a value indicates moving to the left (numbers decrease to the left). For example, $-3 - 2$ is displayed by starting at -3 and moving to the left 2 spaces, if the number line is in increments of 1. This will result in an answer of -5.

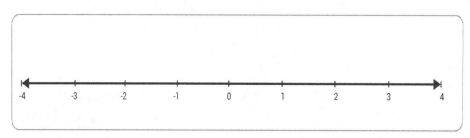

## Rectangular Arrays and Area Models

Rectangular arrays include an arrangement of rows and columns that correspond to the factors and display product totals.

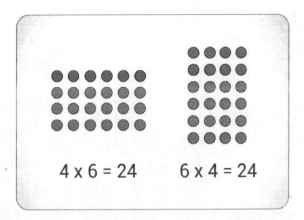

$$4 \times 6 = 24 \qquad 6 \times 4 = 24$$

An area model is a rectangle that is divided into rows and columns that match up to the number of place values within each number. For example, $29 \times 65 = 25 + 4$ and $66 = 60 + 5$. The products of those 4 numbers are found within the rectangle and then summed up to get the answer. The entire process is:

$$(60 \times 25) + (5 \times 25) + (60 \times 4) + (5 \times 4)$$

$$1{,}500 + 240 + 125 + 20 = 1{,}885$$

Here is the actual area model:

|  | **25** | **4** |
|---|---|---|
| **60** | 60x25 <br> 1,500 | 60x4 <br> 240 |
| **5** | 5x25 <br> 125 | 5x4 <br> 20 |

```
    1 , 5 0 0
        2 4 0
        1 2 5
  +        2 0
  ─────────────
    1 , 8 8 5
```

# ACT Math Practice Test #1

1. $3.4 + 2.35 + 4 =$
   - a. 5.35
   - b. 9.2
   - c. 9.75
   - d. 10.25
   - e. 11.15

2. $5.88 \times 3.2 =$
   - a. 18.816
   - b. 16.44
   - c. 20.352
   - d. 17
   - e. 19.25

3. $\frac{3}{25} =$
   - a. 0.15
   - b. 0.1
   - c. 0.9
   - d. 0.75
   - e. 0.12

4. Which of the following is largest?
   - a. 0.45
   - b. 0.096
   - c. 0.3
   - d. 0.313
   - e. 0.25

5. Which of the following is NOT a way to write 40 percent of $N$?
   - a. $(0.4)N$
   - b. $\frac{2}{5}N$
   - c. $40N$
   - d. $\frac{4N}{10}$
   - e. $\frac{8N}{20}$

6. Which is closest to $17.8 \times 9.9$?
   - a. 140
   - b. 180
   - c. 200
   - d. 350
   - e. 400

7. A student gets an 85% on a test with 20 questions. How many answers did the student solve correctly?

    a. 15

    b. 16

    c. 17

    d. 18

    e. 19

8. Four people split a bill. The first person pays for $\frac{1}{5}$, the second person pays for $\frac{1}{4}$, and the third person pays for $\frac{1}{3}$. What fraction of the bill does the fourth person pay?

    a. $\frac{13}{60}$

    b. $\frac{47}{60}$

    c. $\frac{1}{4}$

    d. $\frac{4}{15}$

    e. $\frac{1}{2}$

9. 6 is 30% of what number?

    a. 18

    b. 20

    c. 24

    d. 26

    e. 22

10. $3\frac{2}{3} - 1\frac{4}{5} =$

    a. $1\frac{13}{20}$

    b. $\frac{14}{15}$

    c. $2\frac{2}{3}$

    d. $\frac{4}{5}$

    e. $1\frac{13}{15}$

11. What is $\frac{420}{98}$ rounded to the nearest integer?

    a. 4

    b. 3

    c. 5

    d. 6

    e. 2

12. $4\frac{1}{3} + 3\frac{3}{4} =$

    a. $6\frac{5}{12}$

    b. $8\frac{1}{12}$

    c. $8\frac{2}{3}$

    d. $7\frac{7}{12}$

    e. $9\frac{1}{12}$

13. Five of six numbers have a sum of 25. The average of all six numbers is 6. What is the sixth number?
    a. 8
    b. 10
    c. 11
    d. 12
    e. 13

14. $52.3 \times 10^{-3} =$
    a. 0.00523
    b. 0.0523
    c. 0.523
    d. 523
    e. 5,230

15. If $\frac{5}{2} \div \frac{1}{3} = n$, then $n$ is between:
    a. 5 and 7
    b. 1 and 3
    c. 9 and 11
    d. 3 and 5
    e. 7 and 9

16. A closet is filled with red, blue, and green shirts. If $\frac{1}{3}$ of the shirts are green and $\frac{2}{5}$ are red, what fraction of the shirts are blue?

    a. $\frac{4}{15}$

    b. $\frac{1}{5}$

    c. $\frac{7}{15}$

    d. $\frac{1}{2}$

    e. $\frac{1}{3}$

17. Shawna buys $2\frac{1}{2}$ gallons of paint. If she uses $\frac{1}{3}$ of it on the first day, how much does she have left?

    a. $1\frac{5}{6}$ gallons

    b. $1\frac{1}{2}$ gallons

    c. $1\frac{2}{3}$ gallons

    d. 2 gallons

    e. $2\frac{2}{3}$ gallons

18. If $6t + 4 = 16$, what is $t$?

    a. 1
    b. 2
    c. 3
    d. 4
    e. 5

19. There are $4x + 1$ treats in each party favor bag. If a total of $60x + 15$ treats is distributed, how many bags are given out?

    a. 15
    b. 16
    c. 20
    d. 22
    e. 24

20. In an office, there are 50 workers. A total of 60% of the workers are women, and the chances of a woman wearing a skirt is 50%. If no men wear skirts, how many workers are wearing skirts?

    a. 12
    b. 15
    c. 16
    d. 20
    e. 21

21. If $2x + 6 = 20$, what is $x$?

    a. 3
    b. 4
    c. 7
    d. 9
    e. 11

22. The variable $y$ is directly proportional to $x$. If $y = 3$ when $x = 5$, then what is $y$ when $x = 20$?

    a. 10
    b. 12
    c. 14
    d. 16
    e. 18

23. Apples cost $2 each, while oranges cost $3 each. Maria purchased 10 fruits in total and spent $22. How many apples did she buy?
    a. 5
    b. 6
    c. 7
    d. 4
    e. 8

24. What are the polynomial roots of $x^2 + x - 2$?
    a. 1 and -2
    b. -1 and 2
    c. 2 and -2
    d. 9 and 13
    e. 3 and 9

25. What is the *y*-intercept of $y = x^{5/3} + (x - 3)(x + 1)$?
    a. 3.5
    b. 7.6
    c. -3
    d. -15.1
    e. 3

26. Suppose $\frac{x+2}{x} = 2$. What is *x*?
    a. -1
    b. 0
    c. 2
    d. 4
    e. -2

27. A rectangle has a length that is 5 feet longer than three times its width. If the perimeter is 90 feet, what is the length in feet?
    a. 10
    b. 20
    c. 25
    d. 35
    e. 40

28. Five students take a test. The scores of the first four students are 80, 85, 75, and 60. If the median score is 80, which of the following could NOT be the score of the fifth student?
    a. 60
    b. 80
    c. 85
    d. 100
    e. 90

29. Change $3\frac{3}{5}$ to a decimal.
    a. 3.6
    b. 4.67
    c. 5.3
    d. 0.28
    e. 6.3

30. If a car can travel 300 miles in 4 hours, how far can it go in an hour and a half?
    a. 100 miles
    b. 112.5 miles
    c. 135.5 miles
    d. 150 miles
    e. 155 miles

31. Which measure for the center of a small sample set is most affected by outliers?
    a. Mean
    b. Median
    c. Mode
    d. Range
    e. None of these

32. Given the value of a given stock at monthly intervals, which graph should be used to best represent the trend of the stock?
    a. Box plot
    b. Line plot
    c. Scatter plot
    d. Circle graph
    e. Line graph

33. What is the probability of randomly picking the winner and runner-up from a race of 4 horses and distinguishing which is the winner?
    a. $\frac{1}{4}$
    b. $\frac{1}{2}$
    c. $\frac{1}{16}$
    d. $\frac{1}{12}$
    e. $\frac{1}{3}$

34. Which of the following could be used in the classroom to show $\frac{3}{7} < \frac{5}{6}$ is a true statement?
    a. A bar graph
    b. A number line
    c. An area model
    d. Base 10 blocks
    e. A line graph

35. Add $103{,}678 + 487$
    a. 103,191
    b. 103,550
    c. 104,265
    d. 104,165
    e. 105,143

36. Add $1.001 + 5.629$
    a. 4.472
    b. 4.628
    c. 5.630
    d. 6.245
    e. 6.630

37. Add $143.77 + 5.2$
    a. 138.57
    b. 148.97
    c. 138.97
    d. 148.57
    e. 149.087

38. What is the next number in the following series: $1, 3, 6, 10, 15, 21, \ldots$ ?
    a. 26
    b. 27
    c. 28
    d. 29
    e. 30

39. Add and express in reduced form $\frac{14}{33} + \frac{10}{11}$.
    a. $\frac{2}{11}$

    b. $\frac{6}{11}$

    c. $\frac{4}{3}$

    d. $\frac{44}{33}$

    e. $\frac{3}{4}$

40. 32 is 25% of what number?
    a. 64
    b. 128
    c. 12.65
    d. 8
    e. 155

41. Subtract $112,076 - 1,243$.
    a. 109,398
    b. 113,319
    c. 113,833
    d. 110,833
    e. 111,485

42. Carey bought 184 pounds of fertilizer to use on her lawn. Each segment of her lawn required $11\frac{1}{2}$ pounds of fertilizer to do a sufficient job. If a student were asked to determine how many segments could be fertilized with the amount purchased, what operation would be necessary to solve this problem?
    a. Multiplication
    b. Division
    c. Addition
    d. Subtraction
    e. Exponents

43. Subtract $701.1 - 52.33$.
    a. 753.43
    b. 648.77
    c. 652.77
    d. 638.43
    e. 786.34

44. Which of the following expressions best exemplifies the additive and subtractive identity?
    a. $5 + 2 - 0 = 5 + 2 + 0$
    b. $6 + x = 6 - 6$
    c. $9 - 9 = 0$
    d. $8 + 2 = 10$
    e. $7 + 8 = 15$

45. Which of the following is an equivalent measurement for 1.3 cm?
    a. 0.13 m
    b. 13 m
    c. 0.13 mm
    d. 0.013 mm
    e. 0.013 m

46. Divide $1,015 \div 1.4$.
    a. 7,250
    b. 0.725
    c. 7.25
    d. 725
    e. 575

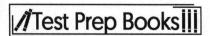

47. At the store, Jan spends $90 on apples and oranges. Apples cost $1 each and oranges cost $2 each. If Jan buys the same number of apples as oranges, how many oranges did she buy?
    a. 20
    b. 25
    c. 30
    d. 35
    e. 40

48. Multiply $12.4 \times 0.2$.
    a. 12.6
    b. 2.48
    c. 12.48
    d. 2.6
    e. 4.28

49. Multiply $1,987 \times 0.05$.
    a. 9.935
    b. 99.35
    c. 993.5
    d. 999.35
    e. 93.95

50. A clock reads 5:00 am. What is the measure of the angle formed by the two hands of that clock?
    a. 300 degrees
    b. 150 degrees
    c. 75 degrees
    d. 210 degrees
    e. 285 degrees

51. Divide, express with a remainder $1,202 \div 44$.
    a. $27\frac{2}{7}$

    b. $2\frac{7}{22}$

    c. $7\frac{2}{7}$

    d. $27\frac{7}{22}$

    e. $22\frac{7}{22}$

52. What is the volume of a box with rectangular sides 5 feet long, 6 feet wide, and 3 feet high?
    a. 60 cubic feet
    b. 75 cubic feet
    c. 80 cubic feet
    d. 14 cubic feet
    e. 90 cubic feet

53. Divide $702 \div 2.6$.
 a. 27
 b. 207
 c. 2.7
 d. 270
 e. 2700

54. A train traveling 50 miles per hour takes a trip lasting 3 hours. If a map has a scale of 1 inch per 10 miles, how many inches apart are the train's starting point and ending point on the map?
 a. 14
 b. 12
 c. 13
 d. 15
 e. 16

55. A traveler takes an hour to drive to a museum, spends 3 hours and 30 minutes there, and takes half an hour to drive home. What percentage of his or her time was spent driving?
 a. 15%
 b. 30%
 c. 40%
 d. 60%
 e. 70%

56. A truck is carrying three cylindrical barrels. Their bases have a diameter of 2 feet, and they have a height of 3 feet. What is the total volume of the three barrels in cubic feet?
 a. $3\pi$
 b. $9\pi$
 c. $12\pi$
 d. $15\pi$
 e. $18\pi$

57. Greg buys a $10 lunch with 5% sales tax. He leaves a $2 tip after his bill. How much money does he spend?
 a. $12.50
 b. $12
 c. $13
 d. $13.25
 e. $13.50

58. Add $5,089 + 10,323$
 a. 15,402
 b. 15,412
 c. 5,234
 d. 15,234
 e. 16,001

59. A teacher is showing students how to evaluate $5 \times 6 + 4 \div 2 - 1$. Which operation should be completed first?
    a. Multiplication
    b. Addition
    c. Division
    d. Subtraction
    e. Exponent

60. What is the definition of a factor of the number 36?
    a. A number that can be divided by 36 and have no remainder
    b. A number that can be added to 36 with no remainder
    c. A prime number that is multiplied times 36
    d. A number that 36 can be divided by and have no remainder
    e. A number that 36 can be multiplied by and have no remainder

# Answer Explanations #1

**1. C:** The decimal points are lined up, with zeroes put in as needed. Then, the numbers are added just like integers:

$$
\begin{array}{r}
3.40 \\
2.35 \\
+4.00 \\
\hline
9.75
\end{array}
$$

**2. A:** This problem can be multiplied as $588 \times 32$, except at the end, the decimal point needs to be moved three places to the left. Performing the multiplication will give 18,816, and moving the decimal place over three places results in 18.816.

**3. E:** The fraction is converted so that the denominator is 100 by multiplying the numerator and denominator by 4, to get $\frac{3}{25} = \frac{12}{100}$. Dividing a number by 100 just moves the decimal point two places to the left, with a result of 0.12.

**4. A:** Figure out which is largest by looking at the first non-zero digits. Choice *B*'s first non-zero digit is in the hundredths place. The other four all have non-zero digits in the tenths place, so it must be *A, C, D,* or *E*. Of these, *A* has the largest first non-zero digit.

**5. C:** $40N$ would be 4000% of *N*. It's possible to check that each of the others is actually 40% of *N*.

**6. B:** Instead of multiplying these out, the product can be estimated by using $18 \times 10 = 180$. The error here should be lower than 15, since it is rounded to the nearest integer, and the numbers add to something less than 30.

**7. C:** 85% of a number means multiplying that number by 0.85. So:

$$
0.85 \times 20 = \frac{85}{100} \times \frac{20}{1}
$$

which can be simplified to:

$$
\frac{17}{20} \times \frac{20}{1} = 17
$$

**8. A:** To find the fraction of the bill that the first three people pay, the fractions need to be added, which means finding the common denominator. The common denominator will be 60.

$$
\frac{1}{5} + \frac{1}{4} + \frac{1}{3} = \frac{12}{60} + \frac{15}{60} + \frac{20}{60} = \frac{47}{60}
$$

The remainder of the bill is:

$$
1 - \frac{47}{60} = \frac{60}{60} - \frac{47}{60} = \frac{13}{60}
$$

**9. B:** 30% is 3/10. The number itself must be 10/3 of 6, or:

$$\frac{10}{3} \times 6 = 10 \times 2 = 20$$

**10. E:** These numbers to improper fractions: $\frac{11}{3} - \frac{9}{5}$. Take 15 as a common denominator:

$$\frac{11}{3} - \frac{9}{5} =: \frac{55}{15} - \frac{27}{15} = \frac{28}{15} = 1\frac{13}{15}$$

.

**11. A:** Dividing by 98 can be approximated by dividing by 100, which would mean shifting the decimal point of the numerator to the left by 2. The result is 4.2 and rounds to 4.

**12. B:** $4\frac{1}{3} + 3\frac{3}{4} = 4 + 3 + \frac{1}{3} + \frac{3}{4} = 7 + \frac{1}{3} + \frac{3}{4}$

Adding the fractions gives:

$$\frac{1}{3} + \frac{3}{4} = \frac{4}{12} + \frac{9}{12} = \frac{13}{12} = 1 + \frac{1}{12}$$

Thus:

$$7 + \frac{1}{3} + \frac{3}{4} = 7 + 1 + \frac{1}{12} = 8\frac{1}{12}$$

**13. C:** The average is calculated by adding all six numbers, then dividing by 6. The first five numbers have a sum of 25. If the total divided by 6 is equal to 6, then the total itself must be 36. The sixth number must be $36 - 25 = 11$.

**14. B:** Multiplying by $10^{-3}$ means moving the decimal point three places to the left, putting in zeroes as necessary.

**15. E:** $\frac{5}{2} \div \frac{1}{3} = \frac{5}{2} \times \frac{3}{1} = \frac{15}{2} = 7.5$.

**16. A:** The total fraction taken up by green and red shirts will be:

$$\frac{1}{3} + \frac{2}{5} = \frac{5}{15} + \frac{6}{15} = \frac{11}{15}$$

The remaining fraction is:

$$1 - \frac{11}{15} = \frac{15}{15} - \frac{11}{15} = \frac{4}{15}$$

**17. C:** If she has used 1/3 of the paint, she has 2/3 remaining. $2\frac{1}{2}$ gallons are the same as $\frac{5}{2}$ gallons.

The calculation is:

$$\frac{2}{3} \times \frac{5}{2} = \frac{5}{3} = 1\frac{2}{3} \text{ gallons}$$

**18. B:** First, subtract 4 from each side. This yields $6t = 12$. Now, divide both sides by 6 to obtain $t = 2$.

**19. A:** Each bag contributes 4x + 1 treats. The total treats will be in the form $4nx + n$ where n is the total number of bags. The total is in the form 60x + 15, from which it is known $n = 15$.

**20. B:** If 60% of 50 workers are women, then there are 30 women working in the office. If half of them are wearing skirts, then that means 15 women wear skirts. Since none of the men wear skirts, this means there are 15 people wearing skirts.

**21. C:** Begin by subtracting 6 from both sides to get $2x = 14$. Dividing both sides by 2 results in $x = 7$.

**22. B:** To be directly proportional means that $y = mx$. If x is changed from 5 to 20, the value of x is multiplied by 4. Applying the same rule to the y-value, also multiply the value of y by 4. Therefore, y = 12.

**23. E:** Let a be the number of apples and o the number of oranges. Then, the total cost is:

$$2a + 3o = 22$$

while it also known that $a + o = 10$. Using the knowledge of systems of equations, cancel the o variables by multiplying the second equation by -3.

This makes the equation:

$$-3a - 3o = -30$$

Adding this to the first equation, the b values cancel to get $-a = -8$, which simplifies to a = 8.

**24. A:** Finding the roots means finding the values of x when y is zero. The quadratic formula could be used, but in this case it is possible to factor by hand, since the numbers -1 and 2 add to 1 and multiply to -2. So, factor:

$$x^2 + x - 2 = (x - 1)(x + 2) = 0$$

then set each factor equal to zero. Solving for each value gives the values x = 1 and x = -2.

**25. C:** To find the y-intercept, substitute zero for x, which gives us:

$$y = 0^{5/3} + (0 - 3)(0 + 1) = 0 + (-3)(1) = -3$$

**26. C:** Multiply both sides by x to get $x + 2 = 2x$, which simplifies to $-x = -2$, or x = 2.

**27. D:** Denote the width as w and the length as $l$. Then, $l = 3w + 5$. The perimeter is $2w + 2l = 90$. Substituting the first expression for $l$ into the second equation yields:

$$2(3w + 5) + 2w = 90$$

$$6w + 10 + 2w = 90$$

$$8w = 80$$

$$w = 10$$

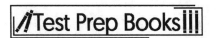

Putting this into the first equation, it yields:

$$l = 3(10) + 5 = 35$$

**28. A:** Lining up the given scores provides the following list: 60, 75, 80, 85, and one unknown. Because the median needs to be 80, it means 80 must be the middle data point out of these five. Therefore, the unknown data point must be the fourth or fifth data point, meaning it must be greater than or equal to 80. The only answer that fails to meet this condition is 60.

**29. A:** 3.6

Divide 3 by 5 to get 0.6 and add that to the whole number 3, to get 3.6. An alternative is to incorporate the whole number 3 earlier on by creating an improper fraction: $\frac{18}{5}$. Then dividing 18 by 5 to get 3.6.

**30. B:** 300 miles in 4 hours is $\frac{300}{4}$ = 75 miles per hour. In 1.5 hours, the car will go $1.5 \times 75$ miles, or 112.5 miles.

**31. A:** Mean. An outlier is a data value that is either far above or far below the majority of values in a sample set. The mean is the average of all the values in the set. In a small sample set, a very high or very low number could drastically change the average of the data points. Outliers will have no more of an effect on the median (the middle value when arranged from lowest to highest) than any other value above or below the median. If the same outlier does not repeat, outliers will have no effect on the mode (value that repeats most often).

**32. E:** Line graph. The scenario involves data consisting of two variables, month and stock value. Box plots display data consisting of values for one variable. Therefore, a box plot is not an appropriate choice. Both line plots and circle graphs are used to display frequencies within categorical data. Neither can be used for the given scenario. Scatter plots compare the values of two variables to see if there are any patterns present. Line graphs display two numerical variables on a coordinate grid and show trends among the variables.

**33. D:** $\frac{1}{12}$. The probability of picking the winner of the race is $\frac{1}{4}$ or:

$$\left( \frac{number\ of\ favorable\ outcomes}{number\ of\ total\ outcomes} \right)$$

Assuming the winner was picked on the first selection, three horses remain from which to choose the runner-up (these are dependent events). Therefore, the probability of picking the runner-up is $\frac{1}{3}$. To determine the probability of multiple events, the probability of each event is multiplied:

$$\frac{1}{4} \times \frac{1}{3} = \frac{1}{12}$$

**34. B:** This inequality can be seen with the use of a number line. $\frac{3}{7}$ is close to $\frac{1}{2}$.

$\frac{5}{6}$ is close to 1, but less than 1, and $\frac{8}{7}$ is greater than 1. Therefore, $\frac{3}{7}$ is less than $\frac{5}{6}$.

**35. D:** 104,165

Set up the problem and add each column, starting on the far right (ones). Add, carrying anything over 9 into the next column to the left. Solve from right to left.

**36. E:** 6.630

Set up the problem, with the larger number on top and numbers lined up at the decimal. Add, carrying anything over 9 into the next column to the left. Solve from right to left.

**37. B:** 148.97

Set up the problem, with the larger number on top and numbers lined up at the decimal. Insert 0 in any blank spots to the right of the decimal as placeholders. Add, carrying anything over 9 into the next column to the left.

**38. C:** Each number in the sequence is adding one more than the difference between the previous two.

For example, $10 - 6 = 4, 4 + 1 = 5$.

Therefore, the next number after 10 is $10 + 5 = 15$.

Going forward, $21 - 15 = 6, 6 + 1 = 7$. The next number is $21 + 7 = 28$.

Therefore, the difference between numbers is the set of whole numbers starting at 2: 2, 3, 4, 5, 6, 7....

**39. C:** $\frac{4}{3}$

Set up the problem and find a common denominator for both fractions:

$$\frac{14}{33} + \frac{10}{11}$$

Multiply each fraction across by 1 to convert to a common denominator:

$$\frac{14}{33} \times \frac{1}{1} + \frac{10}{11} \times \frac{3}{3}$$

Once over the same denominator, add across the top. The total is over the common denominator:

$$\frac{14 + 30}{33} = \frac{44}{33}$$

Reduce by dividing both numerator and denominator by 11:

$$\frac{44 \div 11}{33 \div 11} = \frac{4}{3}$$

**40. B:** 128

This question involves the percent formula:

$$\frac{32}{x} = \frac{25}{100}$$

We multiply the diagonal numbers, 32 and 100, to get 3,200. Dividing by the remaining number, 25, gives us 128.

The percent formula does not have to be used for a question like this. Since 25% is ¼ of 100, you know that 32 needs to be multiplied by 4, which yields 128.

**41. D: 110,833**

Set up the problem, with the larger number on top. Begin subtracting with the far right column (ones). Borrow 10 from the column to the left, when necessary.

**42. B:** This is a division problem because the original amount needs to be split up into equal amounts. The mixed number $11\frac{1}{2}$ should be converted to an improper fraction first:

$$11\frac{1}{2} = \frac{(11 \times 2) + 1}{2} = \frac{23}{2}$$

Carey needs to determine how many times $\frac{23}{2}$ goes into 184. This is a division problem:

$$184 \div \frac{23}{2} = ?$$

The fraction can be flipped, and the problem turns into the multiplication:

$$184 \times \frac{2}{23} = \frac{368}{23}$$

This improper fraction can be simplified into 16 because $368 \div 23 = 16$. The answer is 16 lawn segments.

**43. B:** 648.77

Set up the problem, with the larger number on top and numbers lined up at the decimal. Insert 0 in any blank spots to the right of the decimal as placeholders. Begin subtracting with the far right column. Borrow 10 from the column to the left, when necessary.

**44. A:** The additive and subtractive identity is 0. When added or subtracted to any number, 0 does not change the original number.

**45. E:** 100 cm is equal to 1 m. 1.3 divided by 100 is 0.013. Therefore, 1.3 cm is equal to 0.013 m. Because 1 cm is equal to 10 mm, 1.3 cm is equal to 13 mm.

**46. D:** 725

Set up the division problem.

$$1. \quad 4\overline{)\,1\quad 0\quad 1\quad 5}$$

Move the decimal over one place to the right in both numbers.

$$1\quad 4\overline{)\,1\quad 0\quad 1\quad 5\quad 0}$$

14 does not go into 1 or 10 but does go into 101 so start there.

```
                    7   2   5
  1    4│ 1   0   1   5   0
       -   9   8
               3   5
             -   2   8
                   7   0
                 -   7   0
                       0
```

The result is 725.

**47. C:** One apple/orange pair costs $3 total. Therefore, Jan bought $\frac{90}{3} = 30$ total pairs, and hence, she bought 30 oranges.

**48. B:** 2.48

Set up the problem, with the larger number on top. Multiply as if there are no decimal places. Add the answer rows together. Count the number of decimal places that were in the original numbers ($1 + 1 = 2$).

Place the decimal 2 places to the right for the final solution.

**49. B:** 99.35

Set up the problem, with the larger number on top. Multiply as if there are no decimal places. Add the answer rows together. Count the number of decimal places that were in the original numbers (2).

Place the decimal in that many spots from the right for the final solution.

**50. B:** Each hour on the clock represents 30 degrees. For example, 3:00 represents a right angle. Therefore, 5:00 represents 150 degrees.

**51. D:** $27\frac{7}{22}$

Set up the division problem.

```
  4    4│ 1   2   0   2
```

44 does not go into 1 or 12 but will go into 120 so start there.

```
                    2   7
  4    4│ 1   2   0   2
       -   8   8
               3   2   2
             -   3   0   8
                   1   4
```

The answer is $27\frac{14}{44}$.

Reduce the fraction for the final answer.

$$27\frac{7}{22}$$

**52. E:** The formula for the volume of a box with rectangular sides is the length times width times height, so $5 \times 6 \times 3 = 90$ cubic feet.

**53. D:** 270

Set up the division problem.

$$2.\quad 6\overline{)\;7\quad 0\quad 2\;}$$

Move the decimal over one place to the right in both numbers.

$$2\quad 6\overline{)\;7\quad 0\quad 2\quad 0\;}$$

26 does not go into 7 but does go into 70 so start there.

$$
\begin{array}{r}
2\ 7\ 0 \\
2\ \ 6\overline{)\;7\ \ 0\ \ 2\ \ 0} \\
-\ 5\ \ 2\phantom{\ \ 0\ \ 0} \\
\hline
1\ \ 8\ \ 2\phantom{\ \ 0} \\
-\ 1\ \ 8\ \ 2\phantom{\ \ 0} \\
\hline
0
\end{array}
$$

The result is 270.

**54. D:** First, the train's journey in the real world is $3 \times 50 = 150$ miles. On the map, 1 inch corresponds to 10 miles, so there is $\frac{150}{10} = 15$ inches on the map.

**55. B:** The total trip time is $1 + 3.5 + 0.5 = 5$ hours. The total time driving is $1 + 0.5 = 1.5$ hours. So, the fraction of time spent driving is 1.5/5 or 3/10. To get the percentage, convert this to a fraction out of 100. The numerator and denominator are multiplied by 10, with a result of 30/100. The percentage is the numerator in a fraction out of 100, so 30%.

**56. B:** The formula for the volume of a cylinder is $\pi r^2 h$, where $r$ is the radius and $h$ is the height. The diameter is twice the radius, so these barrels have a radius of 1 foot. That means each barrel has a volume of $\pi \times 1^2 \times 3 = 3\pi$ cubic feet. Since there are three of them, the total is $3 \times 3\pi = 9\pi$ cubic feet.

**57. A:** The tip is not taxed, so he pays 5% tax only on the $10. 5% of $10 is $0.05 \times 10 = \$0.50$. Add up $10 + $2 + $0.50 to get $12.50.

**58. B:** 15,412

Set up the problem and add each column, starting on the far right (ones). Add, carrying anything over 9 into the next column to the left. Solve from right to left.

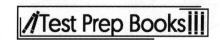

**59. A:** Using the order of operations, multiplication and division are computed first from left to right. Multiplication is on the left; therefore, multiplication should be performed first.

**60. D:** A factor of 36 is any number that can be divided into 36 and have no remainder. $36 = 36 \times 1, 18 \times 2, 9 \times 4,$ and $6 \times 6$. Therefore, it has 7 unique factors: 36, 18, 9, 6, 4, 2, and 1.

# ACT Math Practice Test #2

1. Which of the following numbers has the greatest value?
    a. 1.4378
    b. 1.07548
    c. 1.43592
    d. 0.89409
    e. 0.94739

2. The value of $6 \times 12$ is the same as:
    a. $2 \times 4 \times 4 \times 2$
    b. $7 \times 4 \times 3$
    c. $6 \times 6 \times 3$
    d. $3 \times 3 \times 4 \times 2$
    e. $5 \times 9 \times 8$

3. This chart indicates how many sales of CDs, vinyl records, and MP3 downloads occurred over the last year. Approximately what percentage of the total sales was from CDs?

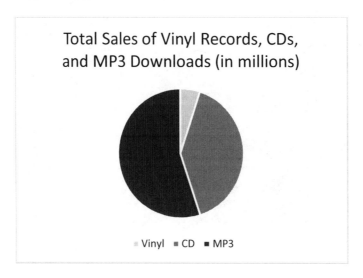

Total Sales of Vinyl Records, CDs, and MP3 Downloads (in millions)

Vinyl   CD   MP3

    a. 55%
    b. 25%
    c. 40%
    d. 5%
    e. 15%

4. Alan currently weighs 200 pounds, but he wants to lose weight to get down to 175 pounds. What is this difference in kilograms? (1 pound is approximately equal to 0.45 kilograms.)
    a. 9 kg
    b. 18.55 kg
    c. 78.75 kg
    d. 90 kg
    e. 11.25 kg

5. Johnny earns $2334.50 from his job each month. He pays $1437 for monthly expenses. Johnny is planning a vacation in 3 months that he estimates will cost $1750 total. How much will Johnny have left over from three months of saving once he pays for his vacation?

    a. $948.50

    b. $584.50

    c. $852.50

    d. $942.50

    e. $742.50

6. Solve the following:

$$(\sqrt{36} \times \sqrt{16}) - 3^2$$

    a. 30

    b. 21

    c. 15

    d. 13

    e. 25

7. In Jim's school, there are 3 girls for every 2 boys. There are 650 students in total. Using this information, how many students are girls?

    a. 260

    b. 130

    c. 65

    d. 390

    e. 410

8. Kimberley earns $10 an hour babysitting, and after 10 p.m., she earns $12 an hour, with the amount paid being rounded to the nearest hour accordingly. On her last job, she worked from 5:30 p.m. to 11 p.m. In total, how much did Kimberley earn on her last job?

    a. $45

    b. $57

    c. $62

    d. $42

    e. $53

9. Arrange the following numbers from least to greatest value:

$$0.85, \frac{4}{5}, \frac{2}{3}, \frac{91}{100}$$

a. $0.85, \frac{4}{5}, \frac{2}{3}, \frac{91}{100}$

b. $\frac{4}{5}, 0.85, \frac{91}{100}, \frac{2}{3}$

c. $\frac{2}{3}, \frac{4}{5}, 0.85, \frac{91}{100}$

d. $0.85, \frac{91}{100}, \frac{4}{5}, \frac{2}{3}$

e. $\frac{4}{5}, \frac{91}{100}, \frac{2}{3}, 0.85$

10. Keith's bakery had 252 customers go through its doors last week. This week, that number increased to 378. Express this increase as a percentage.
   a. 26%
   b. 50%
   c. 35%
   d. 12%
   e. 18%

11. Simplify the following expression:

$$4\frac{2}{3} - 3\frac{4}{9}$$

a. $1\frac{1}{3}$

b. $1\frac{3}{8}$

c. $1$

d. $1\frac{2}{3}$

e. $1\frac{2}{9}$

12. Jessica buys 10 cans of paint. Red paint costs $1 per can and blue paint costs $2 per can. In total, she spends $16. How many red cans did she buy?
   a. 2
   b. 3
   c. 4
   d. 5
   e. 6

13. Six people apply to work for Janice's company, but she only needs four workers. How many different groups of four employees can Janice choose?

    a. 6
    b. 10
    c. 15
    d. 36
    e. 42

14. Which of the following is equivalent to the value of the digit 3 in the number 792.134?

    a. $3 \times 10$

    b. $3 \times 100$

    c. $\frac{3}{10}$

    d. $\frac{3}{100}$

    e. $100 \div 3$

15. In the following expression, which operation should be completed first? $5 \times 6 + (5 + 4) \div 2 - 1.$

    a. Multiplication
    b. Addition
    c. Division
    d. Parentheses
    e. Exponents

16. How will the number 847.89632 be written if rounded to the nearest hundredth?

    a. 847.90
    b. 900
    c. 847.89
    d. 847.896
    e. 850

17. The perimeter of a 6-sided polygon is 56 cm. The length of three of the sides are 9 cm each. The length of two other sides are 8 cm each. What is the length of the missing side?

    a. 11 cm
    b. 12 cm
    c. 13 cm
    d. 10 cm
    e. 9 cm

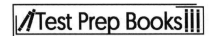

18. Which of the following is a mixed number?

    a. $16\frac{1}{2}$

    b. 16

    c. $\frac{16}{3}$

    d. $\frac{1}{4}$

    e. $\frac{90}{80}$

19. Change 9.3 to a fraction.

    a. $9\frac{3}{7}$

    b. $\frac{903}{1000}$

    c. $\frac{9.03}{100}$

    d. $3\frac{9}{10}$

    e. $9\frac{3}{10}$

20. What is the value of $b$ in this equation?

$$5b - 4 = 2b + 17$$

    a. 13
    b. 24
    c. 7
    d. 21
    e. 15

21. What is 39% of 164?
    a. 63.96
    b. 23.78
    c. 6,396
    d. 2.38
    e. .0987

22. Katie works at a clothing company and sold 192 shirts over the weekend. One third of the shirts that were sold were patterned, and the rest were solid. Which mathematical expression would calculate the number of solid shirts Katie sold over the weekend?

    a. $192 \times \frac{1}{3}$

    b. $192 \div \frac{1}{3}$

    c. $192 \times (1 - \frac{1}{3})$

    d. $192 \div 3$

    e. $192 \times (1 + \frac{1}{3})$

23. Which four-sided shape is always a rectangle?
    a. Rhombus
    b. Square
    c. Parallelogram
    d. Quadrilateral
    e. Trapezoid

24. A rectangle was formed out of pipe cleaner. Its length was $\frac{1}{2}$ ft, and its width was $\frac{11}{2}$ inches. What is its area in square inches?

    a. $\frac{11}{4}$ inch$^2$

    b. $\frac{11}{2}$ inch$^2$

    c. 22 inches$^2$

    d. $27\frac{1}{3}$ inches$^2$

    e. 33 inches$^2$

25. How will $\frac{4}{5}$ be written as a percent?
    a. 40 percent
    b. 125 percent
    c. 90 percent
    d. 80 percent
    e. 25 percent

26. If Danny takes 48 minutes to walk 3 miles, how long should it take him to walk 5 miles maintaining the same speed?
    a. 32 min
    b. 64 min
    c. 80 min
    d. 96 min
    e. 105 min

27. A solution needs 5 ml of saline for every 8 ml of medicine given. How much saline is needed for 45 ml of medicine?

    a. $\frac{225}{8}$ ml

    b. 72 ml

    c. 28 ml

    d. $\frac{45}{8}$ ml

    e. 84 ml

28. What unit of volume is used to describe the following 3-dimensional shape?

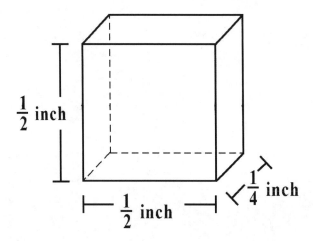

    a. Square inches
    b. Inches
    c. Cubic inches
    d. Squares
    e. Cubic feet

29. Which common denominator would be used in order to evaluate $\frac{2}{3} + \frac{4}{5}$?

    a. 15
    b. 3
    c. 5
    d. 10
    e. 12

30. The diameter of a circle measures 5.75 centimeters. What tool could be used to draw such a circle?

    a. Ruler
    b. Meter stick
    c. Compass
    d. Yard stick
    e. Protractor

31. A piggy bank contains 12 dollars' worth of nickels. A nickel weighs 5 grams, and the empty piggy bank weighs 1050 grams. What is the total weight of the full piggy bank?

    a. 1,110 grams
    b. 1,200 grams
    c. 2,150 grams
    d. 2,250 grams
    e. 2,500 grams

32. Last year, the New York City area received approximately $27\frac{3}{4}$ inches of snow. The Denver area received approximately 3 times as much snow as New York City. How much snow fell in Denver?

    a. $71\frac{3}{4}$ inches

    b. $27\frac{1}{4}$ inches

    c. $89\frac{1}{4}$ inches

    d. $83\frac{1}{4}$ inches

    e. $64\frac{3}{4}$ inches

33. Which of the following would be an instance in which ordinal numbers are used?

    a. Katie scored a 9 out of 10 on her quiz.
    b. Matthew finished second in the spelling bee.
    c. Jacob missed one day of school last month.
    d. Kim was 5 minutes late to school this morning.
    e. Katrina lost 10 lbs. over the summer.

34. How will the following number be written in standard form: $(1 \times 10^4) + (3 \times 10^3) + (7 \times 10^1) + (8 \times 10^0)$

    a. 137
    b. 13,780
    c. 1,378
    d. 13,078
    e. 3,780

35. What is the area of the regular hexagon shown below?

10.39

12

    a. 72
    b. 124.68
    c. 374.04
    d. 748.08
    e. 676.79

36. The area of a given rectangle is 24 square centimeters. If the measure of each side is multiplied by 3, what is the area of the new figure?
    a. 48 cm$^2$
    b. 72 cm$^2$
    c. 111 cm$^2$
    d. 13,824 cm$^2$
    e. 216 cm$^2$

37. Which of the following is the definition of a prime number?
    a. A number that factors only into itself and 1
    b. A number greater than one that factors only into itself and 1
    c. A number less than 10
    d. A number divisible by 10
    e. Any number that's greater than 1

38. Add and express in reduced form $\frac{5}{12} + \frac{4}{9}$
    a. $\frac{9}{17}$
    b. $\frac{1}{3}$
    c. $\frac{31}{36}$
    d. $\frac{3}{5}$
    e. $\frac{15}{29}$

39. Which of the following is the correct order of operations that could be used on a difficult math problem that contained grouping symbols?
    a. Parentheses, Exponents, Multiplication, Division, Addition, Subtraction
    b. Exponents, Parentheses, Multiplication, Division, Addition, Subtraction
    c. Parentheses, Exponents, Addition, Multiplication, Division, Subtraction
    d. Parentheses, Exponents, Division, Addition, Subtraction, Multiplication
    e. Multiplication, Parentheses, Exponents, Division, Subtraction, Addition

40. Convert $\frac{5}{8}$ to a decimal.
    a. 0.62
    b. 1.05
    c. 0.63
    d. 1.60
    e. 1.25

41. Subtract $9,576 - 891$.
    a. 10,467
    b. 9,685
    c. 8,325
    d. 8,685
    e. 7,564

42. If a teacher was showing a class how to round 245.2678 to the nearest thousandth, which place value would be used to decide whether to round up or round down?
    a. Ten-thousandth
    b. Thousandth
    c. Hundredth
    d. Thousand
    e. Hundred

43. Subtract $50.888 - 13.091$.
    a. 63.799
    b. 63.979
    c. 37.979
    d. 37.797
    e. 36.697

44. Students should line up decimal places within the given numbers before performing which of the following?
    a. Multiplication
    b. Division
    c. Subtraction
    d. Exponents
    e. Parentheses

45. Subtract and express in reduced form $\frac{23}{24} - \frac{1}{6}$.
    a. $\frac{22}{18}$
    b. $\frac{11}{9}$
    c. $\frac{19}{24}$
    d. $\frac{4}{5}$
    e. $\frac{1}{2}$

46. Subtract and express in reduced form $\frac{43}{45} - \frac{11}{15}$.

  a. $\frac{10}{45}$

  b. $\frac{16}{15}$

  c. $\frac{32}{30}$

  d. $\frac{5}{9}$

  e. $\frac{2}{9}$

47. Change 0.56 to a fraction.

  a. $\frac{5.6}{100}$

  b. $\frac{14}{25}$

  c. $\frac{56}{1000}$

  d. $\frac{56}{10}$

  e. $\frac{100}{56}$

48. Multiply $13,114 \times 191$.

  a. 2,504,774

  b. 250,477

  c. 150,474

  d. 2,514,774

  e. 1,759,492

49. Marty wishes to save $150 over a 4-day period. How much must Marty save each day on average?

  a. $37.50

  b. $35

  c. $45.50

  d. $41

  e. $43.50

50. A teacher cuts a pie into 6 equal pieces and takes one away. What topic would she be introducing to the class by using such a visual?

  a. Decimals

  b. Addition

  c. Subtraction

  d. Fractions

  e. Exponents

51. Multiply and reduce $\frac{15}{23} \times \frac{54}{127}$.

    a. $\frac{810}{2,921}$

    b. $\frac{81}{292}$

    c. $\frac{69}{150}$

    d. $\frac{810}{2929}$

    e. $\frac{2921}{810}$

52. Which of the following represent one hundred eighty-two billion, thirty-six thousand, four hundred twenty-one and three hundred fifty-six thousandths?

    a. 182,036,421.356
    b. 182,036,421.0356
    c. 182,000,036,421.0356
    d. 182,000,036,421.356
    e. 182,360,000,421.356

53. Divide, express with a remainder $188 \div 16$.

    a. $1\frac{3}{4}$

    b. $111\frac{3}{4}$

    c. $10\frac{3}{4}$

    d. $11\frac{3}{4}$

    e. $3\frac{4}{11}$

54. What other operation could be utilized to teach the process of dividing 9453 by 24 besides division?

    a. Multiplication
    b. Addition
    c. Exponents
    d. Subtraction
    e. Parentheses

55. Bernard can make $80 per day. If he needs to make $300 and only works full days, how many days will this take?

    a. 6
    b. 3
    c. 5
    d. 2
    e. 4

56. A couple buys a house for $150,000. They sell it for $165,000. By what percentage did the house's value increase?
    a. 18%
    b. 13%
    c. 15%
    d. 12%
    e. 10%

57. What operation is repeated to evaluate an expression involving an exponent?
    a. Addition
    b. Multiplication
    c. Division
    d. Subtraction
    e. Parentheses

58. Which of the following formulas would correctly calculate the perimeter of a legal-sized piece of paper that is 14 inches long and $8\frac{1}{2}$ inches wide?
    a. $P = 14 + 8\frac{1}{2}$
    b. $P = 14 + 8\frac{1}{2} + 14 + 8\frac{1}{2}$
    c. $P = 14 \times 8\frac{1}{2}$
    d. $P = 14 \times \frac{17}{2}$
    e. $P = 14 - \frac{17}{2}$

59. Which of the following are units that would be taught in a lecture covering the metric system?
    a. Inches, feet, miles, pounds
    b. Millimeters, centimeters, meters, pounds
    c. Kilograms, grams, kilometers, meters
    d. Teaspoons, tablespoons, ounces
    e. Stone, quart, gallon, foot, yard

60. Which important mathematical property is shown in the expression: $(7 \times 3) \times 2 = 7 \times (3 \times 2)$?
    a. Distributive property
    b. Commutative property
    c. Additive inverse
    d. Associative property
    e. Multiplicative Identity property

# Answer Explanations #2

**1. A:** Compare each numeral after the decimal point to figure out which overall number is greatest. In answers A (1.43785) and C (1.43592), both have the same tenths (4) and hundredths (3). However, the thousandths is greater in answer A (7), so A has the greatest value overall.

**2. D:** By grouping the four numbers in the answer into factors of the two numbers of the question (6 and 12), it can be determined that:

$$(3 \times 2) \times (4 \times 3) = 6 \times 12$$

Alternatively, each of the answer choices could be prime factored or multiplied out and compared to the original value.

$6 \times 12$ has a value of 72 and a prime factorization of $2^3 \times 3^2$.

The answer choices respectively have values of 64, 84, 108, 72, and 360, so Choice D is correct.

**3. C:** The sum total percentage of a pie chart must equal 100%. Since the CD sales take up less than half of the chart and more than a quarter (25%), it can be determined to be 40% overall. This can also be measured with a protractor. The angle of a circle is 360°. Since 25% of 360° would be 90° and 50% would be 180°, the angle percentage of CD sales falls in between; therefore, it would be Choice C.

**4. B:** Using the conversion rate, multiply the projected weight loss of 25 lb. by 0.45 $\frac{kg}{lb}$ to get the amount in kilograms (11.25 kg).

**5. D:** First, subtract $1437 from $2334.50 to find Johnny's monthly savings; this equals $897.50. Then, multiply this amount by 3 to find out how much he will have (in three months) before he pays for his vacation: this equals $2692.50. Finally, subtract the cost of the vacation ($1750) from this amount to find how much Johnny will have left: $942.50.

**6. C:** Follow the *order of operations* in order to solve this problem. Solve the parentheses first, and then follow the remainder as usual:

$$(6 \times 4) - 9$$

This equals $24 - 9$ or 15, answer C.

**7. D:** Three girls for every two boys can be expressed as a ratio: 3:2. This can be visualized as splitting the school into 5 groups: 3 girl groups and 2 boy groups. The number of students which are in each group can be found by dividing the total number of students by 5:

$$\frac{650 \text{ students}}{5 \text{ groups}} = \frac{130 \text{ students}}{\text{group}}$$

To find the total number of girls, multiply the number of students per group (130) by the number of girl groups in the school (3). This equals 390, Choice D.

**8. C:** Kimberley worked 4.5 hours at the rate of $10/h and 1 hour at the rate of $12/h. The problem states that her pay is rounded to the nearest hour, so the 4.5 hours would round up to 5 hours at the rate of $10/h:

$$(5h)(\$10/h) + (1h)(\$12/h) = \$50 + \$12 = \$62$$

**9. C:** The first step is to depict each number using decimals. $\frac{91}{100} = 0.91$

Dividing the numerator by denominator of $\frac{4}{5}$ to convert it to a decimal yields 0.80, while $\frac{2}{3}$ becomes 0.66 recurring. Rearrange each expression in ascending order, as found in Choice *C*.

**10. B:** First, calculate the difference between the larger value and the smaller value:

$$378 - 252 = 126$$

To calculate this difference as a percentage of the original value, and thus calculate the percentage *increase*, divide 126 by 252, then multiply by 100 to reach the percentage 50%, Choice *B*.

**11. E:** Simplify each mixed number of the problem into a fraction by multiplying the denominator by the whole number and adding the numerator:

$$\frac{14}{3} - \frac{31}{9}$$

Since the first denominator is a multiple of the second, simplify it further by multiplying both the numerator and denominator of the first expression by 3 so that the denominators of the fractions are equal:

$$\frac{42}{9} - \frac{31}{9} = \frac{11}{9}$$

Simplifying this further, divide the numerator 11 by the denominator 9; this leaves 1 with a remainder of 2. To write this as a mixed number, place the remainder over the denominator, resulting in $1\frac{2}{9}$.

**12. C:** We are trying to find $x$, the number of red cans. The equation can be set up like this:

$$x + 2(10 - x) = 16$$

The left $x$ is actually multiplied by $1, the price per red can. Since we know Jessica bought 10 total cans, $10 - x$ is the number blue cans that she bought. We multiply the number of blue cans by $2, the price per blue can.

That should all equal $16, the total amount of money that Jessica spent. Working that out gives us:

$$x + 20 - 2x = 16$$

$$20 - x = 16$$

$$x = 4$$

**13. C:** Janice will be choosing 4 employees out of a set of 6 applicants, so this will be given by the choice function. The following equation shows the choice function worked out:

$$\binom{6}{4} = \frac{6!}{4!\,(6-4)!} = \frac{6!}{4!\,(2)!}$$

$$\frac{6 \times 5 \times 4 \times 3 \times 2 \times 1}{4 \times 3 \times 2 \times 1 \times 2 \times 1} = \frac{6 \times 5}{2} = 15$$

**14. D:** $\frac{3}{100}$. Each digit to the left of the decimal point represents a higher multiple of 10 and each digit to the right of the decimal point represents a quotient of a higher multiple of 10 for the divisor.

The first digit to the right of the decimal point is equal to the value $\div$ 10. The second digit to the right of the decimal point is equal to the value $\div (10 \times 10)$, or the value $\div 100$.

**15. D:** Using the order of operations, items inside of parentheses are sorted out first. Therefore, the teacher should resolve the parentheses first. In this expression, multiplication and division are computed next from left to right, followed by addition and subtraction.

**16. A:** 847.90.

The hundredths place value is located two digits to the right of the decimal point (the digit 9 in the original number). The digit to the right of the place value is examined to decide whether to round up or keep the digit. In this case, the digit 6 is 5 or greater so the hundredth place is rounded up. When rounding up, if the digit to be increased is a 9, the digit to its left is increased by one and the digit in the desired place value is made a zero. Therefore, the number is rounded to 847.90.

**17. C:** Perimeter is found by calculating the sum of all sides of the polygon.

$9 + 9 + 9 + 8 + 8 + s = 56$, where $s$ is the missing side length. Therefore, 43 plus the missing side length is equal to 56. The missing side length is 13 cm.

**18. A:** $16\frac{1}{2}$. A mixed number contains both a whole number and either a fraction or a decimal. Therefore, the mixed number is $16\frac{1}{2}$.

**19. E:** $9\frac{3}{10}$. To convert a decimal to a fraction, remember that any number to the left of the decimal point will be a whole number. Then, since 0.3 goes to the tenths place, it can be placed over 10.

**20. C:** To solve for the value of $b$, both sides of the equation need to be equalized.

Start by cancelling out the lower value of -4 by adding 4 to both sides:

$$5b - 4 = 2b + 17$$
$$5b - 4 + 4 = 2b + 17 + 4$$
$$5b = 2b + 21$$

The variable $b$ is the same on each side, so subtract the lower 2b from each side:

$$5b = 2b + 21$$
$$5b - 2b = 2b + 21 - 2b$$
$$3b = 21$$

Then divide both sides by 3 to get the value of $b$:

$$3b = 21$$

$$\frac{3b}{3} = \frac{21}{3}$$

$$b = 7$$

**21. A:** 63.96

This question involves the percent formula. Since we're beginning with a percent, also known as a number over 100, we'll put 39 on the right side of the equation:

$$\frac{x}{164} = \frac{39}{100}$$

Now, multiply 164 and 39 to get 6,396, which then needs to be divided by 100.

$$6,396 \div 100 = 63.96$$

**22. C:** $\frac{1}{3}$ of the shirts sold were patterned. Therefore, $1 - \frac{1}{3} = \frac{2}{3}$ of the shirts sold were solid. Anytime "of" a quantity appears in a word problem, multiplication needs to be used. Therefore, $192 \times \frac{2}{3} =$ $192 \times \frac{2}{3} = \frac{384}{3} = 128$ solid shirts were sold. The entire expression is:

$$192 \times \left(1 - \frac{1}{3}\right)$$

**23. B:** A rectangle is a specific type of parallelogram. It has 4 right angles. A square is a rhombus that has 4 right angles. Therefore, a square is always a rectangle because it has two sets of parallel lines and 4 right angles.

**24. E:** Recall the formula for area, area = length × width. The answer must be in square inches, so all values must be converted to inches. Half of a foot is equal to 6 inches. Therefore, the area of the rectangle is equal to:

$$6 \text{ in} \times \frac{11}{2} \text{ in} = \frac{66}{2} \text{ in}^2 = 33 \text{ in}^2$$

**25. D:** 80 percent. To convert a fraction to a percent, the fraction is first converted to a decimal. To do so, the numerator is divided by the denominator: $4 \div 5 = 0.8$. To convert a decimal to a percent, the number is multiplied by 100: $0.8 \times 10 = 80\%$.

**26. C:** 80 min. To solve the problem, a proportion is written consisting of ratios comparing distance and time. One way to set up the proportion is:

$$\frac{3}{48} = \frac{5}{x} \left(\frac{distance}{time} = \frac{distance}{time}\right)$$

where $x$ represents the unknown value of time. To solve a proportion, the ratios are cross-multiplied:

$$(3)(x) = (5)(48) \rightarrow 3x = 240$$

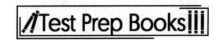

The equation is solved by isolating the variable, or dividing by 3 on both sides, to produce $x = 80$.

**27. A:** Every 8 ml of medicine requires 5 ml. The 45 ml first needs to be split into portions of 8 ml. This results in $\frac{45}{8}$ portions. Each portion requires 5 ml. Therefore:

$$\frac{45}{8} \times 5 = 45 \times \frac{5}{8} = \frac{225}{8} \text{ ml}$$

**28. C:** Volume of this 3-dimensional figure is calculated using length x width x height. Each measure of length is in inches. Therefore, the answer would be labeled in cubic inches.

**29. A:** A common denominator must be found. The least common denominator is 15 because it has both 5 and 3 as factors. The fractions must be rewritten using 15 as the denominator.

**30. C:** A compass is a tool that can be used to draw a circle. The compass would be drawn by using the length of the radius, which is half of the diameter.

**31. D:** A dollar contains 20 nickels. Therefore, if there are 12 dollars' worth of nickels, there are:

$$12 \times 20 = 240 \text{ nickels}$$

Each nickel weighs 5 grams. Therefore, the weight of the nickels is $240 \times 5 = 1{,}200$ grams.

Adding in the weight of the empty piggy bank, the filled bank weighs 2,250 grams.

**32. D:** To find Denver's total snowfall, 3 must be multiplied times $27\frac{3}{4}$. In order to easily do this, the mixed number should be converted into an improper fraction.

$$27\frac{3}{4} = \frac{27 \times 4 + 3}{4} = \frac{111}{4}$$

Therefore, Denver had approximately $\frac{3 \times 111}{4} = \frac{333}{4}$ inches of snow. The improper fraction can be converted back into a mixed number through division.

$$\frac{333}{4} = 83\frac{1}{4} \text{ inches}$$

**33. B:** Ordinal numbers represent a ranking. Placing second in a competition is a ranking among the other participants of the spelling bee.

**34. D:** 13,078. The power of 10 by which a digit is multiplied corresponds with the number of zeros following the digit when expressing its value in standard form. Therefore:

$$(1 \times 10^4) + (3 \times 10^3) + (7 \times 10^1) + (8 \times 10^0)$$

$$10{,}000 + 3{,}000 + 70 + 8 = 13{,}078$$

**35. C:** 374.04. The formula for finding the area of a regular polygon is $A = \frac{1}{2} \times a \times P$ where $a$ is the length of the apothem (from the center to any side at a right angle), and P is the perimeter of the figure. The apothem $a$ is given as 10.39, and the perimeter can be found by multiplying the length of one side

by the number of sides (since the polygon is regular): $P = 12 \times 6 \rightarrow P = 72$. To find the area, substitute the values for $a$ and $P$ into the formula

$$A = \frac{1}{2} \times a \times P$$

$$A = \frac{1}{2} \times (10.39) \times (72) \rightarrow A = 374.04$$

**36. E:** 216 cm². Because area is a two-dimensional measurement, the dimensions are multiplied by a scale that is squared to determine the scale of the corresponding areas. The dimensions of the rectangle are multiplied by a scale of 3. Therefore, the area is multiplied by a scale of $3^2$ (which is equal to 9):

$$24 \, cm^2 \times 9 = 216 \, cm^2$$

**37. B:** A number is prime because its only factors are itself and 1. Positive numbers (greater than one) can be prime numbers.

**38. C:** $\frac{31}{36}$

Set up the problem and find a common denominator for both fractions.

$$\frac{5}{12} + \frac{4}{9}$$

Multiply each fraction across by 1 to convert to a common denominator.

$$\frac{5}{12} \times \frac{3}{3} + \frac{4}{9} \times \frac{4}{4}$$

Once over the same denominator, add across the top. The total is over the common denominator.

$$\frac{15 + 16}{36} = \frac{31}{36}$$

**39. A:** Order of operations follows PEMDAS—Parentheses, Exponents, Multiplication and Division from left to right, and Addition and Subtraction from left to right.

**40. C:** 0.63

Divide 5 by 8, which results in 0.63.

**41. D:** 8,685

Set up the problem, with the larger number on top. Begin subtracting with the far-right column (ones). Borrow 10 from the column to the left, when necessary.

**42. A:** The place value to the right of the thousandth place, which would be the ten-thousandth place, is what gets used. The value in the thousandth place is 7. The number in the place value to its right is greater than 4, so the 7 gets bumped up to 8. Everything to its right turns to a zero, to get 245.2680. The zero is dropped because it is part of the decimal.

**43. D:** 37.797

Set up the problem, larger number on top and numbers lined up at the decimal. Begin subtracting with the far-right column. Borrow 10 from the column to the left, when necessary.

**44. C:** Numbers should be lined up by decimal places before subtraction is performed. This is because subtraction is performed within each place value. The other operations, such as multiplication, division, and exponents (which is a form of multiplication), involve ignoring the decimal places at first and then including them at the end.

**45. C:** $\frac{19}{24}$

Set up the problem and find a common denominator for both fractions.

$$\frac{23}{24} - \frac{1}{6}$$

Multiply each fraction across by 1 to convert to a common denominator.

$$\frac{23}{24} \times \frac{1}{1} - \frac{1}{6} \times \frac{4}{4}$$

Once over the same denominator, subtract across the top.

$$\frac{23 - 4}{24} = \frac{19}{24}$$

**46. E:** $\frac{2}{9}$

Set up the problem and find a common denominator for both fractions.

$$\frac{43}{45} - \frac{11}{15}$$

Multiply each fraction across by 1 to convert to a common denominator.

$$\frac{43}{45} \times \frac{1}{1} - \frac{11}{15} \times \frac{3}{3}$$

Once over the same denominator, subtract across the top.

$$\frac{43 - 33}{45} = \frac{10}{45}$$

Reduce.

$$\frac{10 \div 5}{45 \div 5} = \frac{2}{9}$$

**47. B:** $\frac{14}{25}$

Since 0.56 goes to the hundredths place, it can be placed over 100:

$$\frac{56}{100}$$

Essentially, the way we got there is by multiplying the numerator and denominator by 100:

$$\frac{0.56}{1} \times \frac{100}{100} = \frac{56}{100}$$

Then, the fraction can be simplified down to $\frac{14}{25}$:

$$\frac{56}{100} \div \frac{4}{4} = \frac{14}{25}$$

**48. A:** 2,504,774

Line up the numbers (the number with the most digits on top) to multiply. Begin with the right column on top and the right column on bottom.

Move one column left on top and multiply by the far-right column on the bottom. Remember to add the carry over after you multiply. Continue that pattern for each of the numbers on the top row.

Starting on the far-right column on top repeat this pattern for the next number left on the bottom. Write the answers below the first line of answers; remember to begin with a zero placeholder. Continue for each number in the top row.

Starting on the far-right column on top, repeat this pattern for the next number left on the bottom. Write the answers below the first line of answers. Remember to begin with zero placeholders.

Once completed, ensure the answer rows are lined up correctly, then add.

**49. A:** The first step is to divide up $150 into four equal parts. $\frac{150}{4}$ is 37.5, so she needs to save an average of $37.50 per day.

**50. D:** The teacher would be introducing fractions. If a pie was cut into 6 pieces, each piece would represent $\frac{1}{6}$ of the pie. If one piece was taken away, $\frac{5}{6}$ of the pie would be left over.

**51. A:** $\frac{810}{2921}$

Line up the fractions.

$$\frac{15}{23} \times \frac{54}{127}$$

Multiply across the top and across the bottom:

$$\frac{15 \times 54}{23 \times 127} = \frac{810}{2921}$$

**52. D:** There are no millions, so the millions period consists of all zeros. 182 is in the billions period, 36 is in the thousands period, 421 is in the hundreds period, and 356 is the decimal.

**53. D:** $11\frac{3}{4}$

Set up the division problem.

$$\begin{array}{c|ccc} 1 & 6 & 1 & 8 & 8 \end{array}$$

16 does not go into 1 but does go into 18 so start there.

$$\begin{array}{c|cccc} & & & 1 & 1 \\ 1 & 6 & 1 & 8 & 8 \\ & - & 1 & 6 & \\ \hline & & 2 & 8 \\ & - & 1 & 6 \\ \hline & & 1 & 2 \end{array}$$

The result is $11\frac{12}{16}$

Reduce the fraction for the final answer.

$$11\frac{3}{4}$$

**54. D:** Division can be computed as a repetition of subtraction problems by subtracting multiples of 24.

**55. E:** The number of days can be found by taking the total amount Bernard needs to make and dividing it by the amount he earns per day:

$$\frac{300}{80} = \frac{30}{8} = \frac{15}{4} = 3.75$$

But Bernard is only working full days, so he will need to work 4 days, since 3 days is not a sufficient amount of time.

**56. E:** The value went up by:

$$\$165,000 - \$150,000 = \$15,000$$

Out of \$150,000, this is $\frac{15,000}{150,000} = \frac{1}{10}$. Convert this to having a denominator of 100, the result is $\frac{10}{100}$ or 10%.

**57. B:** A number raised to an exponent is a compressed form of multiplication.

For example, $10^3 = 10 \times 10 \times 10$.

**58. B:** The perimeter of a rectangle is the sum of all four sides.

Therefore, the answer is:

$$P = 14 + 8\frac{1}{2} + 14 + 8\frac{1}{2}$$

$$14 + 14 + 8 + \frac{1}{2} + 8 + \frac{1}{2} = 45 \text{ square inches}$$

**59. C:** Kilograms, grams, kilometers, and meters. Inches, pounds, and baking measurements, such as tablespoons, are not part of the metric system. Kilograms, grams, kilometers, and meters are part of the metric system.

**60. D:** It shows the associative property of multiplication. The order of multiplication does not matter, and the grouping symbols do not change the final result once the expression is evaluated.

# ACT Math Practice Test #3

1. At the beginning of the day, Xavier has 20 apples. At lunch, he meets his sister Emma and gives her half of his apples. After lunch, he stops by his neighbor Jim's house and gives him 6 of his apples. He then uses $\frac{3}{4}$ of his remaining apples to make an apple pie for dessert at dinner. At the end of the day, how many apples does Xavier have left?

    a. 4
    b. 6
    c. 2
    d. 1
    e. 3

2. What is the product of two irrational numbers?
    a. Irrational
    b. Rational
    c. Irrational or rational
    d. Complex and imaginary
    e. Imaginary

3. The graph shows the position of a car over a 10-second time interval. Which of the following is the correct interpretation of the graph for the interval 1 to 3 seconds?

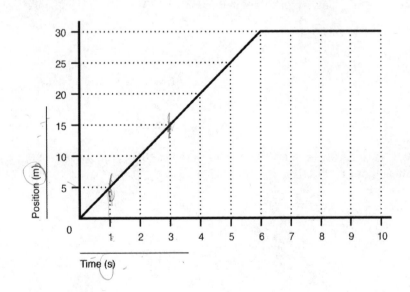

    a. The car remains in the same position.
    b. The car is traveling at a speed of 5 m/s.
    c. The car is traveling up a hill.
    d. The car is traveling at 5 mph.
    e. The car accelerates at a rate of 5 m/s.

4. Being as specific as possible, how is the number -4 classified?
    a. Real, rational, integer, whole, natural
    b. Real, rational, integer, natural
    c. Real, rational, integer
    d. Real, irrational, complex
    e. Real, irrational, whole

5. After a 20% discount, Frank purchased a new refrigerator for $850. How much did he save from the original price?
    a. $170
    b. $212.50
    c. $105.75
    d. $200
    e. $187.50

6. Store brand coffee beans cost $1.23 per pound. A local coffee bean roaster charges $1.98 per 1 ½ pounds. How much more would 5 pounds from the local roaster cost than 5 pounds of the store brand?
    a. $0.55
    b. $1.55
    c. $1.45
    d. $0.45
    e. $0.35

7. What is the solution to the following problem in decimal form?
$$\frac{3}{5} \times \frac{7}{10} \div \frac{1}{2}$$
    a. 0.042
    b. 84%
    c. 0.84
    d. 0.42
    e. 42%

8. Dwayne has received the following scores on his math tests: 78, 92, 83, and 97. What score must Dwayne get on his next math test to have an overall average of 90?
    a. 89
    b. 98
    c. 95
    d. 94
    e. 100

9. What are all the factors of 12?
    a. 12, 24, 36
    b. 1, 2, 4, 6, 12
    c. 12, 24, 36, 48
    d. 1, 2, 3, 4, 6, 12
    e. 0, 1, 12

10. Which of the following augmented matrices represents the system of equations below?

$$2x - 3y + z = -5$$
$$4x - y - 2z = -7$$
$$-x + 2z = -1$$

a. $\begin{bmatrix} 2 & -3 & 1 & 5 \\ 4 & -1 & 0 & -7 \\ -1 & 0 & 2 & 1 \end{bmatrix}$

b. $\begin{bmatrix} 2 & 4 & -1 \\ -3 & -1 & 0 \\ 1 & -2 & 2 \\ -5 & -7 & -1 \end{bmatrix}$

c. $\begin{bmatrix} 2 & 4 & -1 & -5 \\ -3 & -1 & 0 & -7 \\ 2 & -2 & 2 & -1 \end{bmatrix}$

d. $\begin{bmatrix} 2 & -3 & 1 \\ 4 & -1 & -2 \\ -1 & 0 & 2 \end{bmatrix}$

e. $\begin{bmatrix} 2 & -3 & 1 & -5 \\ 4 & -1 & -2 & -7 \\ -1 & 0 & 2 & -1 \end{bmatrix}$

11. What are the zeros of the function: $f(x) = x^3 + 4x^2 + 4x$?
   a. -2
   b. 0, -2
   c. 2
   d. 0, 2
   e. 0

12. If $g(x) = x^3 - 3x^2 - 2x + 6$ and $f(x) = 2$, then what is $g(f(x))$?
   a. -26
   b. 6
   c. $2x^3 - 6x^2 - 4x + 12$
   d. -2
   e. $2x^2 - 6$

13. What is the solution to the following system of equations?
$$x^2 - 2x + y = 8$$
$$x - y = -2$$

   a. $(-2, 3)$
   b. There is no solution.
   c. $(-2, 0)\ (1, 3)$
   d. $(-2, 0)\ (3, 5)$
   e. $(2, 0)\ (-1, 3)$

14. Which of the following is the result after simplifying the expression: $(7n + 3n^3 + 3) + (8n + 5n^3 + 2n^4)$?

    a. $9n^4 + 15n - 2$
    b. $2n^4 + 5n^3 + 15n - 2$
    c. $9n^4 + 8n^3 + 15n$
    d. $2n^4 + 8n^3 + 15n + 3$
    e. $2n^4 + 5n^3 + 15n - 3$

15. What is the product of the following expression?
$$(4x - 8)(5x^2 + x + 6)$$

    a. $20x^3 - 36x^2 + 16x - 48$
    b. $6x^3 - 41x^2 + 12x + 15$
    c. $20x^4 + 11x^2 - 37x - 12$
    d. $2x^3 - 11x^2 - 32x + 20$
    e. $20x^3 - 40x^2 + 24x - 48$

16. How could the following equation be factored to find the zeros?
$$y = x^3 - 3x^2 - 4x$$

    a. $0 = x^2(x - 4), x = 0, 4$
    b. $0 = 3x(x + 1)(x + 4), x = 0, -1, -4$
    c. $0 = x(x + 1)(x + 6), x = 0, -1, -6$
    d. $0 = 3x(x + 1)(x - 4), x = 0, 1, -4$
    e. $0 = x(x + 1)(x - 4), x = 0, -1, 4$

17. The hospital has a nurse-to-patient ratio of 1:25. If there is a maximum of 325 patients admitted at a time, how many nurses are there?

    a. 13 nurses
    b. 25 nurses
    c. 325 nurses
    d. 12 nurses
    e. 5 nurses

18. Which of the following is the solution for the given equation?
$$\frac{x^2 + x - 30}{x - 5} = 11$$

    a. $x = -6$
    b. All real numbers.
    c. $x = 16$
    d. $x = 5$
    e. There is no solution.

19. Mom's car drove 72 miles in 90 minutes. How fast did she drive in feet per second?
    a. 0.8 feet per second
    b. 48.9 feet per second
    c. 0.009 feet per second
    d. 70.4 feet per second
    e. 21.3 feet per second

20. Solve $V = lwh$ for $h$.
    a. $lwV = h$
    b. $h = \dfrac{V}{lw}$
    c. $h = \dfrac{Vl}{w}$
    d. $h = \dfrac{Vw}{l}$
    e. $h = \dfrac{Vl}{w}$

21. What is the domain for the function $y = \sqrt{x}$?
    a. All real numbers
    b. $x \geq 0$
    c. $x > 0$
    d. $y \geq 0$
    e. $x < 0$

22. If Sarah reads at an average rate of 21 pages in four nights, how long will it take her to read 140 pages?
    a. 6 nights
    b. 26 nights
    c. 8 nights
    d. 27 nights
    e. 21 nights

23. The phone bill is calculated each month using the equation $c = 50g + 75$. The cost of the phone bill per month is represented by $c$, and $g$ represents the gigabytes of data used that month. Identify and interpret the slope of this equation.
    a. 75 dollars per day
    b. 75 gigabytes per day
    c. 50 dollars per day
    d. 50 dollars per gigabyte
    e. The slope cannot be determined

24. What is the function that forms an equivalent graph to $y = \cos(x)$?
    a. $y = \tan(x)$
    b. $y = \csc(x)$
    c. $y = \sin\left(x + \dfrac{\pi}{2}\right)$
    d. $y = \sin\left(x - \dfrac{\pi}{2}\right)$
    e. $y = \tan\left(x + \dfrac{\pi}{2}\right)$

25. If $\sqrt{1 + x} = 4$, what is $x$?
    a. 10
    b. 15
    c. 20
    d. 25
    e. 36

26. What is the inverse of the function $f(x) = 3x - 5$?
    a. $f^{-1}(x) = \frac{x}{3} + 5$
    b. $f^{-1}(x) = \frac{5x}{3}$
    c. $f^{-1}(x) = 3x + 5$
    d. $f^{-1}(x) = \frac{x+5}{3}$
    e. $f^{-1}(x) = \frac{x}{3} - 5$

27. What are the zeros of $f(x) = x^2 + 4$?
    a. $x = -4$
    b. $x = \pm 2i$
    c. $x = \pm 2$
    d. $x = \pm 4i$
    e. $x = 2, 4$

28. Twenty is 40% of what number?
    a. 60
    b. 8
    c. 200
    d. 70
    e. 50

29. What is the simplified form of the expression $1.2 * 10^{12} \div 3.0 * 10^8$?
    a. $0.4 * 10^4$
    b. $4.0 * 10^4$
    c. $4.0 * 10^3$
    d. $3.6 * 10^{20}$
    e. $4.0 * 10^2$

30. You measure the width of your door to be 36 inches. The true width of the door is 35.75 inches. What is the relative error in your measurement?
    a. 0.7%
    b. 0.007%
    c. 0.99%
    d. 0.1%
    e. 7.0%

31. What is the y-intercept for $y = x^2 + 3x - 4$?
    a. $y = 1$
    b. $y = -4$
    c. $y = 3$
    d. $y = 4$
    e. $y = -3$

32. Is the following function even, odd, neither, or both?
$$y = \frac{1}{2}x^4 + 2x^2 - 6$$

    a. Even
    b. Odd
    c. Neither
    d. Both
    e. Even for all negative x-values and odd for all positive x-values

33. Which equation is not a function?
    a. $y = |x|$
    b. $y = \sqrt{x}$
    c. $x = 3$
    d. $y = 4$
    e. $y = 3x$

34. How could the following function be rewritten to identify the zeros?
$$y = 3x^3 + 3x^2 - 18x$$

    a. $y = 3x(x + 3)(x - 2)$
    b. $y = x(x - 2)(x + 3)$
    c. $y = 3x(x - 3)(x + 2)$
    d. $y = (x + 3)(x - 2)$
    e. $y = 3x(x + 3)(x + 2)$

35. A six-sided die is rolled. What is the probability that the roll is 1 or 2?
    a. $\frac{1}{6}$
    b. $\frac{1}{4}$
    c. $\frac{1}{3}$
    d. $\frac{1}{2}$
    e. $\frac{1}{36}$

36. A line passes through the origin and through the point (-3, 4). What is the slope of the line?

    a. $-\frac{4}{3}$

    b. $-\frac{3}{4}$

    c. $\frac{4}{3}$

    d. $\frac{3}{4}$

    e. $\frac{1}{3}$

37. What type of function is modeled by the values in the following table?

| $x$ | $f(x)$ |
|---|---|
| 1 | 2 |
| 2 | 4 |
| 3 | 8 |
| 4 | 16 |
| 5 | 32 |

    a. Linear
    b. Exponential
    c. Quadratic
    d. Cubic
    e. Logarithmic

38. An investment of $2,000 is made into an account with an annual interest rate of 5%, compounded continuously. What is the total value of the investment after eight years?

    a. $4,707
    b. $3,000
    c. $2,983.65
    d. $10,919.63
    e. $1,977.61

39. A ball is drawn at random from a ball pit containing 8 red balls, 7 yellow balls, 6 green balls, and 5 purple balls. What's the probability that the ball drawn is yellow?

    a. $^1/_{26}$

    b. $^{19}/_{26}$

    c. $^{14}/_{26}$

    d. 1

    e. $^7/_{26}$

40. Two cards are drawn from a shuffled deck of 52 cards. What's the probability that both cards are Kings if the first card isn't replaced after it's drawn?

    a. $^1/_{169}$

b. $\frac{1}{221}$

c. $\frac{1}{13}$

d. $\frac{4}{13}$

e. $\frac{1}{104}$

41. What's the probability of rolling a 6 at least once in two rolls of a die?

 a. $\frac{1}{3}$

 b. $\frac{1}{36}$

 c. $\frac{1}{6}$

 d. $\frac{1}{12}$

 e. $\frac{11}{36}$

42. Given the set $A = \{1, 2, 3, 4, 5, 6, 7, 8, 9, 10\}$ and $B = \{1, 2, 3, 4, 5\}$, what is $A - (A \cap B)$?

 a. $\{6, 7, 8, 9, 10\}$

 b. $\{1, 2, 3, 4, 5\}$

 c. $\{1, 2, 3, 4, 5, 6, 7, 8, 9, 10\}$

 d. $\emptyset$

 e. $\{-1, -2, -3, -4, -5\}$

43. An equilateral triangle has a perimeter of 18 feet. If a square whose sides have the same length as one side of the triangle is built, what will be the area of the square?

 a. 6 square feet

 b. 36 square feet

 c. 256 square feet

 d. 1000 square feet

 e. 324 square feet

44. In a group of 20 men, the median weight is 180 pounds and the range is 30 pounds. If each man gains 10 pounds, which of the following would be true?

 a. The median weight will increase, and the range will remain the same.

 b. The median weight and range will both remain the same.

 c. The median weight will stay the same, and the range will increase.

 d. The median weight and range will both increase.

 e. The median weight will increase, and the range will decrease.

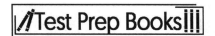

45. For the following similar triangles, what are the values of $x$ and $y$ (rounded to one decimal place)?

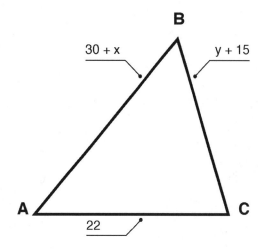

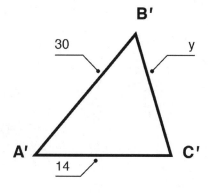

  a. $x = 16.5, y = 25.1$
  b. $x = 19.5, y = 24.1$
  c. $x = 17.1, y = 26.3$
  d. $x = 26.3, y = 17.1$
  e. $x = 24.1, y = 19.5$

46. On Monday, Robert mopped the floor in 4 hours. On Tuesday, he did it in 3 hours. If on Monday, his average rate of mopping was $p$ sq. ft. per hour, what was his average rate on Tuesday?

  a. $\frac{4}{3}p$ sq. ft. per hour
  b. $\frac{3}{4}p$ sq. ft. per hour
  c. $\frac{5}{4}p$ sq. ft. per hour
  d. $p + 1$ sq. ft. per hour
  e. $\frac{1}{3}p$ sq. ft. per hour

47. Which of the following inequalities is equivalent to $3 - \frac{1}{2}x \geq 2$?

  a. $x \geq 2$
  b. $x \leq 2$
  c. $x \geq 1$
  d. $x \leq 1$
  e. $x \leq -2$

48. For which of the following are $x = 4$ and $x = -4$ solutions?

  a. $x^2 + 16 = 0$
  b. $x^2 + 4x - 4 = 0$
  c. $x^2 - 2x - 2 = 0$
  d. $x^2 - x - 16 = 0$
  e. $x^2 - 16 = 0$

49. What are the center and radius of a circle with equation $4x^2 + 4y^2 - 16x - 24y + 51 = 0$?
    a. Center (3, 2) and radius ½
    b. Center (2, 3) and radius ½
    c. Center (3, 2) and radius ¼
    d. Center (2, 3) and radius ¼
    e. Center (2, 2) and radius ¼

50. If the ordered pair $(-3, -4)$ is reflected over the $x$-axis, what's the new ordered pair?
    a. $(-3, -4)$
    b. $(3, -4)$
    c. $(3, 4)$
    d. $(-3, 4)$
    e. $(-4, -3)$

51. If the volume of a sphere is $288\pi$ cubic meters, what are the radius and surface area of the same sphere?
    a. Radius is 6 meters and surface area is $144\pi$ square meters
    b. Radius is 36 meters and surface area is $144\pi$ square meters
    c. Radius is 6 meters and surface area is $12\pi$ square meters
    d. Radius is 36 meters and surface area is $12\pi$ square meters
    e. Radius 12 meters and surface area $144\pi$ square meters

52. The triangle shown below is a right triangle. What's the value of $x$?

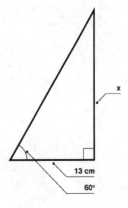

    a. $x = 1.73$
    b. $x = 0.57$
    c. $x = 13$
    d. $x = 14.73$
    e. $x = 22.52$

53. Ten students take a test. Five students get a 50. Four students get a 70. If the average score is 55, what was the last student's score?
    a. 20
    b. 40
    c. 50
    d. 60
    e. 62

54. A sample data set contains the following values: 1, 3, 5, 7. What's the standard deviation of the set?
   a. 2.58
   b. 4
   c. 6.23
   d. 1.1
   e. 0.25

55. A company invests $50,000 in a building where they can produce saws. If the cost of producing one saw is $40, then which function expresses the amount of money the company pays? The variable $y$ is the money paid and $x$ is the number of saws produced.
   a. $y = 50,000x + 40$
   b. $y + 40 = x - 50,000$
   c. $y = 40x - 50,000$
   d. $y = 40x + 50,000$
   e. $y = 4x + 50,000$

56. A pair of dice is thrown, and the sum of the two scores is calculated. What's the expected value of the roll?
   a. 5
   b. 6
   c. 7
   d. 8
   e. 9

57. A line passes through the point (1, 2) and crosses the $y$-axis at $y = 1$. Which of the following is an equation for this line?
   a. $y = 2x$
   b. $y = x + 1$
   c. $x + y = 1$
   d. $y = \frac{x}{2} - 2$
   e. $y = x - 1$

58. $x^4 - 16$ can be simplified to which of the following?
   a. $(x^2 - 4)(x^2 + 4)$
   b. $(x^2 + 4)(x^2 + 4)$
   c. $(x^2 - 4)(x^2 - 4)$
   d. $(x^2 - 2)(x^2 + 4)$
   e. $(x^2 - 2)(x^2 + 2)$

59. $(4x^2y^4)^{\frac{3}{2}}$ can be simplified to which of the following?

   a. $8x^3y^6$

   b. $4x^{\frac{5}{2}}y$

   c. $4xy$

   d. $32x^{\frac{7}{2}}y^{\frac{11}{2}}$

   e. $x^3y^6$

60. A ball is thrown from the top of a high hill, so that the height of the ball as a function of time is $h(t) = -16t^2 + 4t + 6$, in feet. What is the maximum height of the ball in feet?

   a. 6
   b. 6.25
   c. 6.5
   d. 6.75
   e. 6.8

# Answer Explanations #3

**1. D:** This problem can be solved using basic arithmetic. Xavier starts with 20 apples, then gives his sister half, so 20 divided by 2.

$$\frac{20}{2} = 10$$

He then gives his neighbor 6, so 6 is subtracted from 10.

$$10 - 6 = 4$$

Lastly, he uses ¾ of his apples to make an apple pie, so to find remaining apples, the first step is to subtract ¾ from one and then multiply the difference by 4.

$$\left(1 - \frac{3}{4}\right) \times 4 = ?$$

$$\left(\frac{4}{4} - \frac{3}{4}\right) \times 4 = ?$$

$$\left(\frac{1}{4}\right) \times 4 = 1$$

**2. C:** The product of two irrational numbers can be rational or irrational. Sometimes, the irrational parts of the two numbers cancel each other out, leaving a rational number. For example, $\sqrt{2} \times \sqrt{2} = 2$ because the roots cancel each other out. Technically, the product of two irrational numbers can be complex because complex numbers can have either the real or imaginary part (in this case, the imaginary part) equal zero and still be considered a complex number. However, Choice *D* is incorrect because the product of two irrational numbers is not an imaginary number so saying the product is complex *and* imaginary is incorrect.

**3. B:** The car is traveling at a speed of five meters per second. On the interval from one to three seconds, the position changes by ten meters. By making this change in position over time into a rate, the speed becomes ten meters in two seconds or five meters in one second.

**4. C:** The number negative four is classified as a real number because it exists and is not imaginary. It is rational because it does not have a decimal that never ends. It is an integer because it does not have a fractional component. The next classification would be whole numbers, for which negative four does not qualify because it is negative. Choices *D* and *E* are wrong because -4 is not considered an irrational number because it does not have a never-ending decimal component.

**5. B:** Since $850 is the price *after* a 20% discount, $850 represents 80% of the original price. To determine the original price, set up a proportion with the ratio of the sale price (850) to original price (unknown) equal to the ratio of sale percentage (where $x$ represents the unknown original price):

$$\frac{850}{x} = \frac{80}{100}$$

To solve a proportion, cross multiply the numerators and denominators and set the products equal to each other: $(850)(100) = (80)(x)$. Multiplying each side results in the equation $85,000 = 80x$.

To solve for $x$, divide both sides by 80: $\frac{85,000}{80} = \frac{80x}{80}$, resulting in $x = 1062.5$. Remember that $x$ represents the original price. Subtracting the sale price from the original price ($1062.50 - \$850$) indicates that Frank saved $212.50.

**6. D:** $0.45

List the givens.

$$Store\ coffee = \$1.23/lbs$$

$$Local\ roaster\ coffee = \$1.98/1.5\ lbs$$

Calculate the cost for 5 lbs of store brand.

$$\frac{\$1.23}{1\ lbs} \times 5\ lbs = \$6.15$$

Calculate the cost for 5 lbs of the local roaster.

$$\frac{\$1.98}{1.5\ lbs} \times 5\ lbs = \$6.60$$

Subtract to find the difference in price for 5 lbs.

$$
\begin{array}{r}
\$6.60 \\
-\$6.15 \\
\hline
\$0.45
\end{array}
$$

**7. C:** The first step in solving this problem is expressing the result in fraction form. Separate this problem first by solving the division operation of the last two fractions. When dividing one fraction by another, invert or flip the second fraction and then multiply the numerator and denominator.

$$\frac{7}{10} \times \frac{2}{1} = \frac{14}{10}$$

Next, multiply the first fraction with this value:

$$\frac{3}{5} \times \frac{14}{10} = \frac{42}{50}$$

In this instance, you can find the decimal form by converting the fraction into $\frac{x}{100}$, where $x$ is the number from which the final decimal is found. Multiply both the numerator and denominator by 2 to get the fraction as an expression of $\frac{x}{100}$.

$$\frac{42}{50} \times \frac{2}{2} = \frac{84}{100}$$

In decimal form, this would be expressed as 0.84.

**8. E:** To find the average of a set of values, add the values together and then divide by the total number of values. In this case, include the unknown value of what Dwayne needs to score on his next test, in order to solve it.

$$\frac{78 + 92 + 83 + 97 + x}{5} = 90$$

Add the unknown value to the new average total, which is 5. Then multiply each side by 5 to simplify the equation, resulting in:

$$78 + 92 + 83 + 97 + x = 450$$

$$350 + x = 450$$

$$x = 100$$

Dwayne would need to get a perfect score of 100 in order to get an average of at least 90.

Test this answer by substituting back into the original formula:

$$\frac{78 + 92 + 83 + 97 + 100}{5} = 90$$

**9. D:** 1, 2, 3, 4, 6, 12. A given number divides evenly by each of its factors to produce an integer (no decimals). The number 5, 7, 8, 9, 10, 11 (and their opposites) do not divide evenly into 12. Therefore, these numbers are not factors.

**10. E:** The augmented matrix that represents the system of equations has dimensions $4 \times 3$ because there are three equations with three unknowns. The coefficients of the variables make up the first three columns, and the last column is made up of the numbers to the right of the equals sign. This system can be solved by reducing the matrix to row-echelon form, where the last column gives the solution for the unknown variables.

**11. B:** There are two zeros for the function: $x = 0, -2$.

The zeros can be found several ways, but this particular equation can be factored into:

$$f(x) = x(x^2 + 4x + 4) = x(x + 2)(x + 2)$$

 By setting each factor equal to zero and solving for $x$, there are two solutions. On a graph these zeros can be seen where the line crosses the $x$-axis.

**12. D:** This problem involves a composition function, where one function is plugged into the other function. In this case, the $f(x)$ function is plugged into the $g(x)$ function for each $x$-value. The composition equation becomes:

$$g(f(x)) = 2^3 - 3(2^2) - 2(2) + 6$$

Simplifying the equation gives the answer:

$$g(f(x)) = 8 - 3(4) - 2(2) + 6 = 8 - 12 - 4 + 6 = -2$$

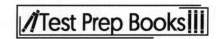

**13. D:** This system of equations involves one quadratic function and one linear function, as seen from the degree of each equation. One way to solve this is through substitution. Solving for y in the second equation yields $y = x + 2$. Plugging this equation in for the y of the quadratic equation yields:

$$x^2 - 2x + x + 2 = 8$$

Simplifying the equation, it becomes:

$$x^2 - x + 2 = 8$$

Setting this equal to zero and factoring, it becomes:

$$x^2 - x - 6 = 0 = (x - 3)(x + 2)$$

Solving these two factors for $x$ gives the zeros $x = 3, -2$. To find the y-value for the point, each number can be plugged in to either original equation. Solving each one for y yields the points $(3, 5)$ and $(-2, 0)$.

**14. D:** The expression is simplified by collecting like terms. Terms with the same variable and exponent are like terms, and their coefficients can be added.

**15. A:** Finding the product means distributing one polynomial onto the other. Each term in the first must be multiplied by each term in the second. Then, like terms can be collected. Multiplying the factors yields the expression:

$$20x^3 + 4x^2 + 24x - 40x^2 - 8x - 48$$

Collecting like terms means adding the $x^2$ terms and adding the $x$ terms. The final answer after simplifying the expression is:

$$20x^3 - 36x^2 + 16x - 48$$

**16. E:** Finding the zeros for a function by factoring is done by setting the equation equal to zero, then completely factoring. Since there was a common $x$ for each term in the provided equation, that would be factored out first. Then the quadratic that was left could be factored into two binomials, which are $(x + 1)(x - 4)$. Setting each factor equal to zero and solving for $x$ yields three zeros.

**17. A:** 13 nurses

Using the given information of 1 nurse to 25 patients and 325 patients, set up an equation to solve for number of nurses (N):

$$\frac{N}{325} = \frac{1}{25}$$

Multiply both sides by 325 to get N by itself on one side.

$$\frac{N}{1} = \frac{325}{25} = 13 \ nurses$$

**18. E:** The equation can be solved by factoring the numerator into $(x + 6)(x - 5)$.

Since that same factor exists on top and bottom, that factor $(x - 5)$ cancels.

This leaves the equation $x + 6 = 11$.

Solving the equation gives the answer $x = 5$. When this value is plugged into the equation, it yields a zero in the denominator of the fraction. Since this is undefined, there is no solution.

**19. D:** This problem can be solved by using unit conversion. The initial units are miles per minute. The final units need to be feet per second. Converting miles to feet uses the equivalence statement 1 mile equals 5,280 feet. Converting minutes to seconds uses the equivalence statement 1 minute equals 60 seconds. Setting up the ratios to convert the units is shown in the following equation:

$$\frac{72 \text{ mi}}{90 \text{ min}} \times \frac{1 \text{ min}}{60 \text{ s}} \times \frac{5280 \text{ ft}}{1 \text{ mi}} = 70.4 \frac{\text{ft}}{\text{s}}$$

The initial units cancel out, and the new units are left.

**20. B:** The formula can be manipulated by dividing both the length, *l*, and the width, *w*, on both sides. The length and width will cancel on the right, leaving height by itself.

**21. B:** The domain is all possible input values, or *x*-values. For this equation, the domain is every number greater than or equal to zero. There are no negative numbers in the domain because taking the square root of a negative number results in an imaginary number.

**22. D:** This problem can be solved by setting up a proportion involving the given information and the unknown value. The proportion is:

$$\frac{21 \text{ } pages}{4 \text{ } nights} = \frac{140 \text{ } pages}{x \text{ } nights}$$

Solving the proportion by cross-multiplying, the equation becomes $21x = 4 \times 140$, where $x = 26.67$. Since it is not an exact number of nights, the answer is rounded up to 27 nights. Twenty-six nights would not give Sarah enough time.

**23. D:** The slope from this equation is 50, and it is interpreted as the cost per gigabyte used. Since the *g*-value represents number of gigabytes and the equation is set equal to the cost in dollars, the slope relates these two values. For every gigabyte used on the phone, the bill goes up 50 dollars.

**24. C:** Graphing the function $y = \cos(x)$ shows that the curve starts at $(0, 1)$, has an amplitude of 2, and a period of $2\pi$. This same curve can be constructed using the sine graph, by shifting the graph to the left $\frac{\pi}{2}$ units. This equation is in the form:

$$y = \sin\left(x + \frac{\pi}{2}\right)$$

**25. B:** Start by squaring both sides to get $1 + x = 16$. Then subtract 1 from both sides to get $x = 15$.

**26. A:** This inverse of a function is found by switching the $x$ and y in the equation and solving for y. In the given equation, solving for y is done by adding 5 to both sides, then dividing both sides by 3. This answer can be checked on the graph by verifying the lines are reflected over $y = x$.

**27. B:** The zeros of this function can be found by using the quadratic formula:

$$x = \frac{-b \pm \sqrt{b^2 - 4ac}}{2a}$$

Identifying $a$, $b$, and $c$ can also be done from the equation because it is in standard form. The formula becomes:

$$x = \frac{0 \pm \sqrt{0^2 - 4(1)(4)}}{2(1)} = \frac{\sqrt{-16}}{2}$$

Since there is a negative underneath the radical, the answer is a complex number:

$$x = \pm 2i$$

**28. E:** Setting up a proportion is the easiest way to represent this situation. The proportion becomes $\frac{20}{x} = \frac{40}{100}$, where cross-multiplication can be used to solve for $x$. The answer can also be found by observing the two fractions as equivalent, knowing that twenty is half of forty, and fifty is half of one-hundred.

**29. C:** Division with scientific notation can be solved by grouping the first terms together and grouping the tens together. The first terms can be divided, and the tens terms can be simplified using the rules for exponents. The initial expression becomes $0.4 * 10^4$. This is not in scientific notation because the first number is not between 1 and 10. Shifting the decimal and subtracting one from the exponent yields $4.0 * 10^3$.

**30. A:** The relative error can be found by finding the absolute error and making it a percent of the true value. The absolute error is $36 - 35.75 = 0.25$. This error is then divided by 36—the true value—to find 0.7%.

**31. B:** The y-intercept of an equation is found where the $x$-value is zero. Plugging zero into the equation for $x$ allows the first two terms to cancel out, leaving -4.

**32. A:** The equation is *even* because:

$$f(-x) = f(x)$$

Plugging in a negative value will result in the same answer as when plugging in the positive of that same value. The function:

$$f(-2) = \frac{1}{2}(-2)^4 + 2(-2)^2 - 6 = 8 + 8 - 6 = 10$$

yields the same value as:

$$f(2) = \frac{1}{2}(2)^4 + 2(2)^2 - 6 = 8 + 8 - 6 = 10$$

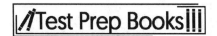

**33. C:** The equation $x = 3$ is not a function because it does not pass the vertical line test. This test is made from the definition of a function, where each $x$-value must be mapped to one, and only one, y-value. This equation is a vertical line, so the $x$-value of 3 is mapped with an infinite number of y-values.

**34. A:** The function can be factored to identify the zeros. First, the term $3x$ is factored out to the front because each term contains $3x$. Then, the quadratic is factored into $(x + 3)(x - 2)$.

**35. C:** A die has an equal chance for each outcome. Since it has six sides, each outcome has a probability of $\frac{1}{6}$. The chance of a 1 or a 2 is therefore $\frac{1}{6} + \frac{1}{6} = \frac{1}{3}$.

**36. A:** The slope is given by:

$$m = \frac{y_2 - y_1}{x_2 - x_1} = \frac{0 - 4}{0 - (-3)} = -\frac{4}{3}$$

**37. B:** The table shows values that are increasing exponentially. The differences between the inputs are the same, while the differences in the outputs are changing by a factor of 2. The values in the table can be modeled by the equation $f(x) = 2^x$.

**38. C:** The formula for continually compounded interest is $A = Pe^{rt}$. Plugging in the given values to find the total amount in the account yields the equation:

$$A = 2000e^{0.05*8} = 2983.65$$

**39. E:** The sample space is made up of $8 + 7 + 6 + 5 = 26$ balls. The probability of pulling each individual ball is $^1/_{26}$. Since there are 7 yellow balls, the probability of pulling a yellow ball is $^7/_{26}$.

**40. B:** For the first card drawn, the probability of a King being pulled is $\frac{4}{52}$. Since this card isn't replaced, if a King is drawn first the probability of a King being drawn second is $\frac{3}{51}$. The probability of a King being drawn in both the first and second draw is the product of the two probabilities: $\frac{4}{52} \times \frac{3}{51} = \frac{12}{2652}$. This fraction, when divided by 12, equals $\frac{1}{221}$.

**41. E:** The addition rule is necessary to determine the probability because a 6 can be rolled on either roll of the die. The rule used is:

$$P(A \text{ or } B) = P(A) + P(B) - P(A \text{ and } B)$$

The probability of a 6 being individually rolled is $\frac{1}{6}$ and the probability of a 6 being rolled twice is:

$$\frac{1}{6} \times \frac{1}{6} = \frac{1}{36}$$

Therefore, the probability that a 6 is rolled at least once is:

$$\frac{1}{6} + \frac{1}{6} - \frac{1}{36} = \frac{11}{36}$$

**42. A:** $(A \cap B)$ is equal to the intersection of the two sets A and B, which is $\{1, 2, 3, 4, 5\}$. $A - (A \cap B)$ is equal to the elements of A that are *not* included in the set $(A \cap B)$. Therefore:

$$A - (A \cap B) = \{6, 7, 8, 9, 10\}$$

**43. B:** An equilateral triangle has three sides of equal length, so if the total perimeter is 18 feet, each side must be 6 feet long. A square with sides of 6 feet will have an area of $6^2 = 36$ square feet.

**44. A:** If each man gains 10 pounds, every original data point will increase by 10 pounds. Therefore, the man with the original median will still have the median value, but that value will increase by 10.

The smallest value and largest value will also increase by 10 and, therefore, the difference between the two won't change. The range does not change in value and, thus, remains the same.

**45. C:** Because the triangles are similar, the lengths of the corresponding sides are proportional. Therefore:

$$\frac{30 + x}{30} = \frac{22}{14} = \frac{y + 15}{y}$$

This results in the equation $14(30 + x) = 22 \times 30$ which, when solved, gives $x = 17.1$.

The proportion also results in the equation $14(y + 15) = 22y$ which, when solved, gives $y = 26.3$.

**46. A:** Robert accomplished his task on Tuesday in ¾ the time compared to Monday. He must have worked 4/3 as fast.

**47. B:** To simplify this inequality, subtract 3 from both sides to get $-\frac{1}{2}x \geq -1$. Then, multiply both sides by -2 (remembering this flips the direction of the inequality) to get $x \leq 2$.

**48. E:** There are two ways to approach this problem. Each value can be substituted into each equation. Choice *A* can be eliminated, since $4^2 + 16 = 32$. Choice *B* can be eliminated, since:

$$4^2 + 4 \cdot 4 - 4 = 28$$

Choice *C* can be eliminated, since:

$$4^2 - 2 \cdot 4 - 2 = 6$$

But, plugging in either value into $x^2 - 16$ gives:

$$(\pm 4)^2 - 16 = 16 - 16 = 0$$

**48. E:** There are two ways to approach this problem. Each value can be substituted into each equation. Choice *A* can be eliminated, since:

$$4^2 + 16 = 32$$

Choice *B* can be eliminated, since:

$$4^2 + 4 \cdot 4 - 4 = 28$$

Choice *C* can be eliminated, since:

$$4^2 - 2 \cdot 4 - 2 = 6$$

But, plugging in either value into $x^2 - 16$ gives:

$$(\pm 4)^2 - 16 = 16 - 16 = 0$$

**49. B:** The technique of completing the square must be used to change:

$$4x^2 + 4y^2 - 16x - 24y + 51 = 0$$

into the standard equation of a circle. First, the constant must be moved to the right-hand side of the equals sign, and each term must be divided by the coefficient of the $x^2$ term (which is 4). The $x$ and $y$ terms must be grouped together to obtain:

$$x^2 - 4x + y^2 - 6y = -\frac{51}{4}$$

Next, the process of completing the square must be completed for each variable. This gives:

$$(x^2 - 4x + 4) + (y^2 - 6y + 9) = -\frac{51}{4} + 4 + 9$$

The equation can be written as:

$$(x - 2)^2 + (y - 3)^2 = \frac{1}{4}$$

Therefore, the center of the circle is (2, 3) and the radius is:

$$\sqrt{1/4} = 1/2$$

**50. D:** When an ordered pair is reflected over an axis, the sign of one of the coordinates must change. When it's reflected over the $x$-axis, the sign of the $y$-coordinate must change. The $x$-value remains the same. Therefore, the new ordered pair is $(-3, 4)$.

**51. A:** Because the volume of the given sphere is $288\pi$ cubic meters, this means $\frac{4}{3}\pi r^3 = 288\pi$. This equation is solved for $r$ to obtain a radius of 6 meters. The formula for the surface area of a sphere is $4\pi r^2$, so if $r = 6$ in this formula, the surface area is $144\pi$ square meters.

**52. E:** SOHCAHTOA is used to find the missing side length. Because the angle and adjacent side are known, $\tan 60 = \frac{x}{13}$. Making sure to evaluate tangent with an argument in degrees, this equation gives:

$$x = 13 \tan 60 = 13 \times \sqrt{3} = 22.52$$

**53. A:** Let the unknown score be $x$. The average will be:

$$\frac{5 \times 50 + 4 \times 70 + x}{10} = \frac{530 + x}{10} = 55$$

Multiply both sides by 10 to get $530 + x = 550$, or $x = 20$.

**54. A:** First, the sample mean must be calculated:

$$\bar{x} = \frac{1}{4}(1 + 3 + 5 + 7) = 4$$

The standard deviation of the data set is $s = \sqrt{\frac{\sum(x-\bar{x})^2}{n-1}}$, and $n = 4$ represents the number of data points. Therefore:

$$s = \sqrt{\frac{1}{3}[(1-4)^2 + (3-4)^2 + (5-4)^2 + (7-4)^2]} = \sqrt{\frac{1}{3}(9+1+1+9)} = 2.58$$

**55. D:** For manufacturing costs, there is a linear relationship between the cost to the company and the number produced, with a $y$-intercept given by the base cost of acquiring the means of production, and a slope given by the cost to produce one unit. In this case, that base cost is $50,000, while the cost per unit is $40. So, $y = 40x + 50,000$.

**56. C:** The expected value is equal to the total sum of each product of individual score and probability. There are 36 possible rolls.

The probability of rolling a 2 is $\frac{1}{36}$.

The probability of rolling a 3 is $\frac{2}{36}$.

The probability of rolling a 4 is $\frac{3}{36}$.

The probability of rolling a 5 is $\frac{4}{36}$.

The probability of rolling a 6 is $\frac{5}{36}$.

The probability of rolling a 7 is $\frac{6}{36}$.

The probability of rolling an 8 is $\frac{5}{36}$.

The probability of rolling a 9 is $\frac{4}{36}$.

The probability of rolling a 10 is $\frac{3}{36}$.

The probability of rolling an 11 is $\frac{2}{36}$.

Finally, the probability of rolling a 12 is $\frac{1}{36}$.

Each possible outcome is multiplied by the probability of it occurring. Like this:

$$2 \times \frac{1}{36} = a$$

$$3 \times \frac{2}{36} = b$$

$$4 \times \frac{3}{36} = c$$

And so forth.

Then all of those results are added together:

$$a + b + c \ldots = expected\ value$$

In this case, it equals 7.

**57. B:** From the slope-intercept form, $y = mx + b$, it is known that $b$ is the $y$-intercept, which is 1. Compute the slope as $\frac{2-1}{1-0} = 1$, so the equation should be $y = x + 1$.

**58. A:** This has the form $t^2 - y^2$, with $t = x^2$ and $y = 4$.

It's also known that:

$$t^2 - y^2 = (t + y)(t - y)$$

Substituting the values for $t$ and $y$ into the right-hand side gives:

$$(x^2 - 4)(x^2 + 4)$$

**59. A:** Simplify this to:

$$(4x^2y^4)^{\frac{3}{2}} = 4^{\frac{3}{2}}(x^2)^{\frac{3}{2}}(y^4)^{\frac{3}{2}}$$

Now:

$$4^{\frac{3}{2}} = (\sqrt{4})^3 = 2^3 = 8$$

For the other, recall that the exponents must be multiplied, so this yields:

$$8x^{2 \cdot \frac{3}{2}}y^{4 \cdot \frac{3}{2}} = 8x^3y^6$$

**60. B:** The independent variable's coordinate at the vertex of a parabola (which is the highest point, when the coefficient of the squared independent variable is negative) is given by $x = -\frac{b}{2a}$. Substitute and solve for $x$ to get:

$$x = -\frac{4}{2(-16)} = \frac{1}{8}$$

Using this value of $x$, the maximum height of the ball ($y$), can be calculated.

Substituting *x* into the equation yields:

$$h(t) = -16\frac{1}{8}^2 + 4\frac{1}{8} + 6 = 6.25$$

Dear ACT Test Taker,

We would like to start by thanking you for purchasing this study guide for your ACT exam. We hope that we exceeded your expectations.

Our goal in creating this study guide was to cover all of the topics that you will see on the test. We also strove to make our practice questions as similar as possible to what you will encounter on test day. With that being said, if you found something that you feel was not up to your standards, please send us an email and let us know.

We would also like to let you know about other books in our catalog that may interest you.

## SAT

This can be found on Amazon: amazon.com/dp/1628456868

## ACCUPLACER

amazon.com/dp/1628459344

## TSI

amazon.com/dp/1628457341

## AP Biology

amazon.com/dp/1628456221

We have study guides in a wide variety of fields. If the one you are looking for isn't listed above, then try searching for it on Amazon or send us an email.

Thanks Again and Happy Testing!
Product Development Team
info@studyguideteam.com

# FREE Test Taking Tips DVD Offer

To help us better serve you, we have developed a Test Taking Tips DVD that we would like to give you for FREE. **This DVD covers world-class test taking tips that you can use to be even more successful when you are taking your test.**

All that we ask is that you email us your feedback about your study guide. Please let us know what you thought about it – whether that is good, bad or indifferent.

To get your **FREE Test Taking Tips DVD**, email freedvd@studyguideteam.com with "FREE DVD" in the subject line and the following information in the body of the email:

> a. The title of your study guide.

> b. Your product rating on a scale of 1-5, with 5 being the highest rating.

> c. Your feedback about the study guide. What did you think of it?

> d. Your full name and shipping address to send your free DVD.

If you have any questions or concerns, please don't hesitate to contact us at freedvd@studyguideteam.com.

Thanks again!